# Figure & Form

VOLUME I

*Skills and Expression*

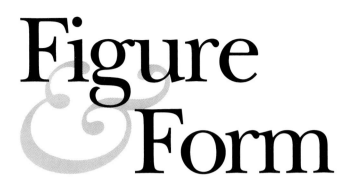

# Figure & Form

VOLUME I

*Skills and Expression*

Lu Bro

WCB Brown &
Benchmark

**Book Team**

Production Editor  *Kay J. Brimeyer*
Designer  *Elise A. Burckhardt*
Photo Editor  *Sharon A. LaPrell*
Art Editor  *Kenneth E. Ley*
Visuals/Design Developmental Consultant  *Marilyn A. Phelps*

**WCB Brown &**
**Benchmark**
A Division of Wm. C. Brown Communications, Inc.

Vice President and General Manager  *Thomas E. Doran*
Executive Managing Editor  *Ed Bartell*
Executive Editor  *Edgar J. Laube*
Director of Marketing  *Kathy Law Laube*
National Sales Manager  *Eric Ziegler*
Marketing Manager  *Kathleen Nietzke*
Advertising Manager  *Jodi Rymer*
Managing Editor, Production  *Colleen A. Yonda*
Manager of Visuals and Design  *Faye M. Schilling*

Design Manager  *Jac Tilton*
Art Manager  *Janice Roerig*
Photo Manager  *Shirley Charley*
Production Editorial Manager  *Ann Fuerste*
Permissions/Records Manager  *Connie Allendorf*

**Wm. C. Brown Communications, Inc.**

Chairman Emeritus  *Wm. C. Brown*
Chairman and Chief Executive Officer  *Mark C. Falb*
President and Chief Operating Officer  *G. Franklin Lewis*
Corporate Vice President, Operations  *Beverly Kolz*
Corporate Vice President, President of WCB Manufacturing  *Roger Meyer*

**Front cover and Section I Opener:** George Kolbe, German, b. 1877. *Female Nude with Right Leg Raised*, pen and washed 47.6 × 30.5 cm. Courtesy of the Art Institute of Chicago.
**Back cover:** Agustus John, "Nude Study." Pencil, 48.3 × 29.2 cm. From Peter Harris Collection. Artist's Source: *Nude to Naked* by Georg Eisler.
**Section II Opener:** Paul Wunderlich. "Adam and Eva I." 1970 Acryl on linen. 162 × 130 cm. (seen in Paul Wunderlich by Jens Christian Jensen. Editor, Herausqeqeban von Volker Huber, p. 83.)

Library of Congress Catalog Card Number: 91–71682

ISBN 0–697–03059–8

Printed in the United States of America by Wm. C. Brown Communications, Inc., 2460 Kerper Boulevard, Dubuque, IA 52001

10  9  8  7  6  5  4  3  2  1

# DEDICATION

▼

Dedicating this book to my husband, Andrew, is my way of honoring a long give and take editorial relationship that is sure to continue even after he discovers what I took from what he gave.

# CONTENTS

▼

▲

# PREFACE

▼

**F**igure and Form, in its two-volume complement, encourages several uses. Instructors in figure drawing are urged to examine both volumes, then select and adapt from this comprehensive study for individual course needs. No two drawing instructors stress drawing problems similarly. Consequently, these volumes are intended to support many approaches and emphases.

The classroom student would be well advised to practice exercises not incorporated into a programmed sequence. In studio courses practice is what effects change. If your instructor emphasizes anatomy, well and good. But alongside, try supplementing what you have learned with other skill development and expressive studies as well. For those students whose teachers by-pass anatomy but who emphasize skill and expressive exploration, your own initiative in drawing bones and musculature would be exceptional groundwork in support of expressive efforts.

Any interested person who wants to self-teach figure drawing would find the sequences in these volumes extremely useful. Individual development should carefully follow the chapter sequencing. Two persons might work together trading modeling chores, alleviating a big expense. But truthfully, the best venue is the classroom for direction, peer support, feedback, and access to a model.

The pedagogical philosophy in these volumes is practice AND play. There is no real substitute for academic exercises, for *practicing the scales* of figure drawing. Though hard work is necessary that way, not everything related to the figure can be done through the eye to the mind to the hand. Allowing the student artist *play time*, to let nonsense take its own direction, is equally critical. These two volumes speak to finding the uniqueness of each person in his or her drawing. We know that skill can take us to the heights of verisimilitude, meaning the appearance of being real. Beginning artists most often want to work realistically

because realism has specific, known ends. However, students should be reminded to learn to let go, to experiment.

Freedom, of course, does not mean total license. Not every drawing stroke is a good one. Sensitive marks carry information. But many students, in their struggles for specific answers, in trying to be correct or "right," ignore the beauty of the inner self, the potential expressive self. Making marks can help a person find his or her own very special gifts in drawing, and there are studies herein to help.

Let me explain how the books are organized. In Volume I, within Section One, exercises follow traditional expectations.

Chapter Two, *The Warm-Ups*, is initiated with gestures for the figure. Some gestures will take thirty seconds, others sixty, others a number of minutes. Gestures with brief, intense studies allow the student new at figure drawing a chance to become comfortable with the nude as a subject. Drawing gesturally for fifteen to twenty minutes as a preliminary warm-up before each class tends to yield more active participation and sustains spontaneity in longer studies. Of course, each of the fourteen gestural concepts can be expanded and used as ends in themselves.

*Rehearsing What is Known About Imaging*, the next chapter, examines shorthand perspective by way of sighting vertically and horizontally. Composition is discussed. Both references and illustrations identify some broader types of esthetic space such as realism, idealism, linear and aerial perspective, and so on.

*Contour-line studies* incorporate blind contour with values, a figure turning in space, sighting methods discussed earlier (looking for angles and intersections of the body), the topographical cross contour, creating compositions with nonsequential forms, foreshortening (a variation from frontal sighting), the fluid use of succinct line, when lines become shapes, and the possibilities of figures transformed into landscapes.

*Value, Mass, and Texture* are emphasized next. Beginning with types of lighting, studies move through cross-hatching, eraser drawings, mirror images with distortions, positives inversing negatives, tonal ranges with different papers, finger drawings, and the use of mixed media.

The last chapter of this section, *Adopt An Artist,* opens new spatial alternatives for the entry level student. Practicing an artist's theory or methodology allows a student to embrace a different way of perceiving space. Most persons attracted to the fine arts work with consistent, undivided perceptions. But consistency sometimes is synonymous with falling into a rut. Stepping into someone else's shoes, holding the pencil of a master artist, can prompt fresh insights, choices, and judgements. The idea is to adopt a theory for a time, use it as one's own, but *not* to copy an artist's work.

In Section Two more concentrated efforts in expressive work are explored. I feel that the figure has been, and continues to be, treated in at least five paradigmatic ways, and that learning to draw the figure expressively can be serviced by the studies nestled within each paradigm. Briefly, the five paradigms are these:

*Figure As Form*—how proportion, idealized beauty, and verisimilitude join. The student can play with ratios, canons, and proportions used by civilizations and individual artists. Often exploration with ratios results in humorous images. One can also learn transferring mechanisms, drawing figures from several points of view with exactitude.

*Figure Against Form*—how the breaking of the Renaissance picture plane loosens the figure from its skin and bones. This paradigm works with the loss of the "ideal," discussed in the first paradigm. Students will read of types and kinds of dissolution; from the heroine and hero to the prostitute and the pimp, using noble and refined methods to abusing drawing and painting surfaces purposefully.

*Figure Above Form*—how an iconographic ordering, either cultural or religious, infers completeness. Study and work with icons indicates something beyond the image itself. Cultural icons might include the sexy woman standing in front of an automobile or the male wrestler. Religious icons stem largely from Buddhist, Christian, American Indian, African, and some Jewish roots.

*Figure and Form, a Paradox*—how the accidental and surreal, the anti-art and the polarization of values achieve a near-mystical bridge among visual ambiguities. Within this paradigm, students work with *Chance,* the playground of the accidental stroke, the possibility of finding something from a chaos of lines. Students are further guided into surrealistic paradoxes as well as the paradoxes in conceptual art.

*Figure, the Fulfiller of Form*—how the human figure services an imagination that lies in narrative or illustrative sources. A visual layout in this paradigm invites students to select from good literature, narrative, history or biography, the characters and scenes, the media and methods, and the final physical form for a work of art. After all, one of our greatest illustrators was Michelangelo. Remember the biblical painted scenes on the Sistine ceiling?

Finally, in *Putting it Together,* we look at twelve students who have put skills and expression together. They share where their inspirations and ideas come from, what media they used, how they used media and materials, and how they knew when their work was finished. Nearly all of these students have moved into graduate work in the fine arts.

Book Two of this two-part set has two sections as well. Section One is *An Anatomical Coloring Book for Artists,* a nuts-and-bolts sequential grouping of anatomical pictures that takes students through the human body part by part, head to toe, surface to bone. The coloring book is a convenient guide to muscles. Free-hand drawing processes are included and are intended to support the practices encouraged in the first volume. There is a long and thorough sequencing of body muscles in motion intended to be used alongside a live model in a studio or classroom. This section concludes with a series of anatomical drawings by master artists.

Section Two of Book Two is titled *People and Parts.* Sooner or later every figurative artist needs a visual glossary, a series of photographs of actual models, male and female, posed similarly but alternating left side to right side. The glossary tries to move past the limitation of having only right or left body positions, forcing the artist to transpose. Maybe some of these photographs will help you when you are caught without reference to a real limb or torso. ▲

# ACKNOWLEDGEMENTS

▼

Textbooks are rare indeed without the forbearance of students. Hundreds could join a list of those supplying the tough questions and reasoned feedback that moved me to my computer to write a second text. The teaching of figurative art is a quixiotic art form in itself. Much information is skills oriented, usually with a specific anatomical thrust. But as much growth, perhaps more, derives from insights about the ways human beings relate to their bodies and to their emotions. Both willing and unwilling wrestling by students eventually personalized the questions of space, of illusion and form, and became an invitation for me to share where I could. Without the wrestling and sharing, this book would not exist.

Three research assistants I gratefully credit for moving the manuscript to completion. One is my daughter, Sarah Bro, a medical researcher whose expertise and pedagogical insights allowed not only creative arrangement for the section on anatomy but a professional consistency. Generously, she believed a job should be well done, and that stretched her into much overtime on my behest.

James Negley, for my purposes a master of computers, set up databases, completed data entries, read proofs, provided many figurative drawings of specific processes, recommended solutions to problems, contacted professional artists for reproductions of their work, assisted in layout, researched in libraries and museums, and with indulgent kindness listened to cycles of woe of this beleaguered author. In truth, the second volume might have expired were it not for the dedication of Sarah Bro and Jim Negley.

Kay-lynne Johnson, the third and final assistant, researched sections of art history, filed, developed macros, created two illustrative icons, contributed computer art detailing her procedures, cut and taped hundreds of images to make illustration sheets, gave general assistance and learned to type in the process, her humor sustaining us both.

Peter Krumhardt was the photographer with whom I worked most closely. His patience and competence made possible the reproductions of a full range of works herein. His sessions included settings both in and out of the classroom. Indeed, the section titled "People and Parts" is all Krumhardt's work. He was tireless, supportive, helpful, even with marginal originals.

David Pavlik, Coordinator of Media Graphics at Iowa State University, lent his considerable analytical and artistic talents toward recreating Medieval, Gothic, Renaissance, and Baroque systems of drawing figures. I am grateful for his very specialized efforts.

Inevitably, one needs the expertise of teaching colleagues. I have leaned on a few. William Berry has been generous with tips, cautions, and knowledge byond his own excellent text, a book I have used for years. John Weinkein, with his specialized interests in American Indians, worked with me on a section related to them. Bob Lorr guided me to several student anatomical works which are in the book. Even more, he shared his experienced concerns about teaching anatomy for artists, a challenging discipline. Stick-with-it support from my department head, Evan Firestone, sustained a long process.

Special thanks go to two extraordinary librarians. Diane Childs and Carlota Cutierrez can find nearly anything, and I sent them often to obscure territory. They sought out and retrieved minutae no one else could. Two-high-energy library "go-fers," Don Rouse and Kevin Hotop, mined data which from exhaustion I would have omitted with great regret.

Apart from the arduous task of writing a text, every book has certain special moments, moments of conviviality and community. From the hundreds of dedicated students I salute the group who, one full afternoon, gave drawing time for the book in exchange for a free model, materials and food: Jim Negley, Pam Shewanik, Kris Lucas, Paul Guy, Pete Hansen, Linda McFadden and Richard Heger.

Katherine Nolan of Fine Art Picture Research in Chicago, was professional and humane in meeting the needs of the text. I found Ms. Nolan exceptionally resourceful in tracking some of the more esoteric images. She relieved my burdens considerably and I will be forever gratefgul.

Deep in the trenches pulling this together under considerable pressure were Meredith Morgan, Editor, Kay Brimeyer, Production Editor, Sharon LaPrell, Photo Editor, Ken Ley, Art Editor and Elise Burckhardt, Designer.

Finally, I must salute one of my heroes, H. Richard Niebuhr. His *Christ and Culture*, Harper & Row, 1951, was pivotal in my own search for bases on which to build the expressive paradigms. After testing several possibilities— an art history chronology, a Jungian psychological foundation, an esthetic or philosophical process—and, when nothing else worked, Niebuhr's paradigms gave me mine. I simply changed the titles. "The Enduring Problem" became "Figure As Form." "Christ Against Culture" became "Figure Against Form." "The Christ Above Culture" became "Figure Above Form." "Christ and Culture in Paradox" became "Figure and Form: A Paradox" and "The Christ of Culture" became "Figure and Form, the Narrative." I stood on the shoulders of this giant and saw with clarity what I should do.

# Figure & Form

VOLUME I

*Skills and Expression*

# ONE

## *Figure and Form*

# 1

▼

# Why Do We Draw the Figure?

"**B**eing is the great explainer," Thoreau said in his journal in 1841. We spend our lives working out our mystery. Our meaning, our truths, our persuasions are subjective but interfaced with everyone else.

Our markings on paper are both discursive and autographic. Everyone confesses something about himself or herself in a drawing, but no one else confesses quite like we artists do. Our body/spirit and mind/emotions rest in the trails we leave, the lines and pictures we draw.

A word of caution: we artists produce work, spectators observe it. We express, the observer is impressed. We do not set out to create works that can be used by spectators to define us, our personalities. Rather we create works to shape feelings, and those feelings are often shared by numbers of people. But no matter what images we draw or paint, our unique and very personal markings are always intrinsic to the end result.

For instance, if we take the figurative subject of *Adam and Eve* and review six variations, we begin to see what exquisite parallels and differences human beings reference.

Masaccio's *The Expulsion from Paradise,* indicates a new realism in showing the nude human body in motion. Both Adam and Eve visibly express shame and grief. There is little visible space, though the foreshortened flying angel suggests recession.

Jan van Eyck's Adam and Eve from *The Ghent Altarpiece* were painted just less than life size. The figures, turned toward the center, are placed at opposite ends of a huge altarpiece, 11′3″ high × 14′5″ wide. Both figures are painted in a slight three-point perspective, which accommodates the viewer's position below, looking up, an innovation at the time. The Northern Renaissance artists, contrary to the Italian Renaissance artists, created female proportions that included narrowly placed breasts and elongated, often swollen, abdomens.

***Figure 1.1*** *Masaccio (1401–1428).* The Expulsion from Paradise, *c. 1427. Fresco. Brancacci Chapel, Sta. Maria del Carmine, Florence. Scala/Art Resource, New York.*

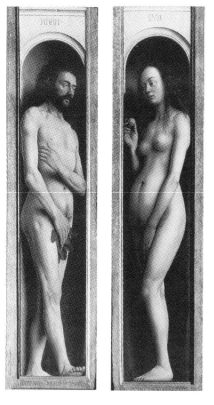

**Figure 1.2** *Jean van Eyck (1395–1441). Adam and Eve. Detail of the Ghent Altarpiece. 1432. Open 11'3" × 14'5". St. Bavo, Ghent. © A. C. L., Bruxelles. Scala/Art Resource, New York.*

**Figure 1.3** *Albrecht Durer (1471–1528). Adam and Eve. Meder 1. 1504. Engraving. (Trimmed to plate mark). 249 × .193 (9¹³⁄₁₆ × 7⅝) National Gallery of Art, Washington. Gift of R. Horace Gallatin B-15, 224.*

Albrecht Durer's *Adam and Eve,* are mathematically constructed figures. This engraving was finished in 1504 when Durer was laboring to find ideal proportions for the human figure. The figures stand here as a means to an end. Durer's end was to build perfect human figures. His means was a rational use of geometric proportions.

Paul Wunderlich's *Adam und Eva I,* painted in 1970, is the reverse of Albrecht Durer's work, almost as though Wunderlich projected a slide and traced the image onto his life-size canvas. Since Durer's work was an engraving, he would have worked on his etching plate with Adam on the right and Eve on the left. The printing process would have reversed the figures. But Wunderlich purposely exhumes Durer's mathematical system for play. He overlays both bodies with geometric shapes and compass points, he gives Eve florescent red hair, he includes the goat, bull, rabbit, cat, mouse, and parrot of Durer's work, painting them in subdued colors but with eerie lighting. Durer's space is idealized nature. Wunderlich's space is supernatural.

Martha Mayer Erlebacher's *Adam and Eve,* 1975, is a diptych, meaning two panel pieces. The earlier van Eyck work was two panels as well, one on either end of a huge altarpiece. Erlebacher's panels are separate yet together. The figures pull away from one another, each holding an apple in one hand. Eve points to Adam as though she received her apple from him. The background is diminished, no leaves or shrubbery cover either crotch. Neither body is idealized. Both male and female look as though the figures were drawn and painted in a naturalistic way.

Paul Hempe's play on the Adam and Eve title reads *Adav and Eem.* His ideas come from two sources, life and pre-Renaissance works of art. He is fond of the quirky, decorative things in illuminated manuscripts and wants to play with those ideas and religious anecdotes by placing the figures in different contexts. In early manuscripts artists found ways to describe yet cover up nudity, such as the placing of masks over the genitals. The mask with the long nose identifies the male, the mask with the wide mouth, the female. "Sharing the fruits of their labor," the words below the image, is not only a pun on the apple, it applies to the circularity of the arms and hands around the fruit, suggesting unison and sharing.

Six people who chose the same subject, Adam and Eve, *expressed* very different things, allowing very distinct *impressions* for viewers. The parallel was the choosing of Adam and Eve. The difference was what each artist showed about the famous primeval couple.

No matter what media or materials—dry, grease, or wet base; no matter what intellectual system—mathematical, sighting, perspective; no matter what expressive approach—realistic, expressionistic, abstract; no matter what scale—large or small; no matter what visual form—drawing, painting, sculpture—drawing the human figure is one way we begin to understand how we fit with others, with nature, and with the universe. ▲

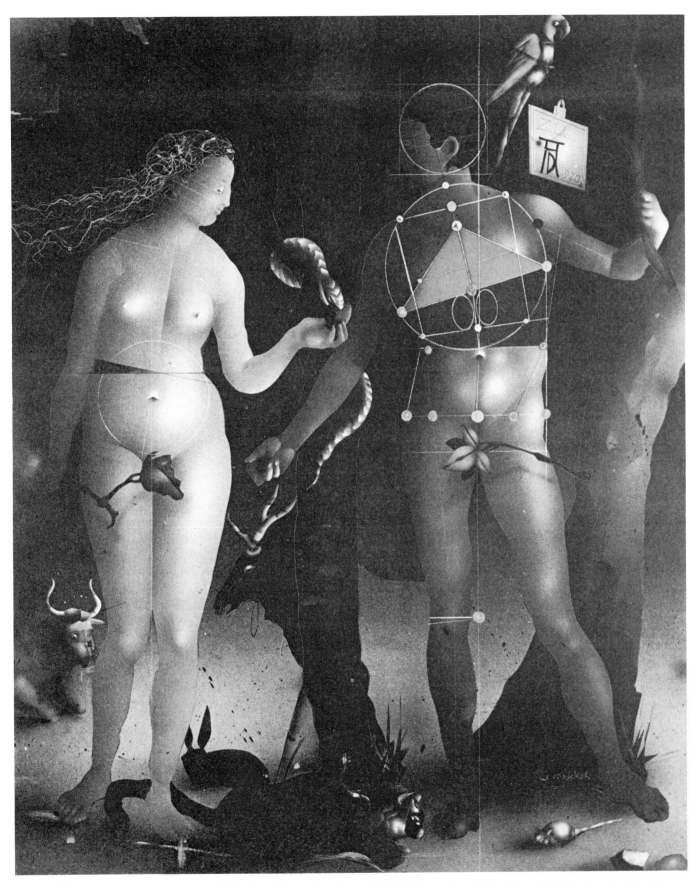

**Figure 1.4** *Paul Wunderlich.* Adam und Eva I. *1970. Acryl on linen. 162 × 130 cm. (seen in* Paul Wunderlich *by Jens Christian Jensen. Editor, Herausgegeban von Volker Huber, pg. 83)*

***Figure 1.5*** *Martha Mayer Erlebacher.* Adam and Eve. *1975. Oil on canvas. 64"h × 40"w. Sloan Fund Purchase. Sloan Collection of American Painting, Valpariso University Museum of Art.*

**Figure 1.6** *Paul Hempe.* Adav and Eem. *1988. Mixed media on rag paper.*

# 2

▼

# Blind Gesture

## Materials

*3B or 4B pencil*
*18″ × 24″ smooth newsprint pad*

For a great many students, gesture is the affective entry to true expressive drawing. All drawing is gesture in the sense that every stroke is a psycho-motor translation. Your personality and your hand combine with your perception. Unfortunately, gesture is difficult to explain with precision. Gesture is a translation of the essential characteristics of those physical forms you see and from which you draw. Gesture captures structural rhythm in space, which means linear flow, which in turn means either static or motive energy, which finally means the kinetic abstraction of a whole form. To add to the difficulty, the learning experience in gesture usually comes after you do it.

Most persons who begin to draw start with what appears to them as shapes with edges. Edges as such do not exist in nature, but an arbitrary edge is often used to identify forms in space. However, if you try to draw a gesture with outer edges, you miss the point of gesture drawing.

In the beginning stages, gesture derives from seeing the structural flow and rhythm of the whole figure. Be aware that gestural lines do not tell all. Detail is not their objective. They imply. They suggest the interlock of mass and weight, the cylinders of the limbs and body. They isolate the linear rhythm of the figure.

Quality in a gesture is elusive. Quality results from seeing weight, density, thrust, force, stress, and time, factors to be discussed as the problems become more complicated. The point is to embody in lines the *essential* characteristics, the structural design, of that whole form you are drawing. Let the characteristics dictate your stroke.

**Figure 2.1**  *Drawing on the outer edges is not gesture drawing.*

**Figure 2.2**  *Gesture showing linear rhythm of a figure.*

**Figure 2.3**  *A gesture of figurative dynamics.*

Your first gestures will vary in quality. In drawing a blind gesture, you will not be able to refer to your paper; your eyes will remain on the figure at all times. Bear in mind with any exercise that asks you to draw with no reference to the paper, you are learning first to *sensitize* your eye to what is before it, and to do so in a very condensed way. Drawing "blind" actually moves you away from preconceived notions about being right or correct. The exercise is also trying to suspend premature notions about space. Most students seem to feel that drawing is done *on* some surface. Rather, good art finally comes from within the total space before you, meaning that the image, materials, strokes, and methods are all arranged in such a way as to shape meaning. These first gestures are a step toward that end.

Figures present complexities and subtleties in perspective, in member interrelationships, in surface variety, and in movement. Gestural drawing is critical to understanding how to represent a figure. For example, to describe the differences in "body English," (the stance of a person) the strokes you make, your lines, need to indicate that person's structural tensions, the counterbalanced units of the mass and any other information your eye can translate. Gesture is important in giving your stroke time to mature, to carry more and more complex information. Drawing literal dimensions, that is, edges, is not the objective. Drawing figurative dynamics is.

A good gesture conveys the expressive mass of a figure in space (masses of the head, torso, upper and lower arms, and legs). Remember, figures do go all the way around. Try to feel them with your eyes and move your pencil with that feeling, like sculpting with lines.

**Figure 2.4** *Blind gesture with structural lines and mass.*

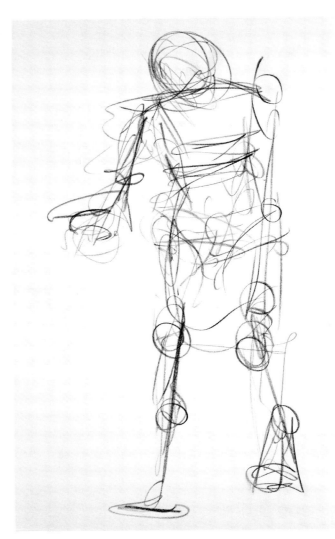

**Figure 2.5** *Gesture indicating structure and mass.*

The illustrations of gestures shown here are individual interpretations. Do not try to copy someone else's translation. Prize your own efforts in seeing the figure through gesture because your unique perceptions are intrinsic to *your* strokes.

You can expect problems to arise. Drawing blind does not leave much control over heights, lengths, and widths. It does not leave much control over making compositions. But those objectives are not what the exercise is asking of you. What it is asking is that you *see*, that you try to see mass and structural flow. Varying the pressure of your pencil will help. Less pressure for areas more pronounced, more pressure for areas receding, as if you were carving on the paper. Your first step will be to interpret the body's major rhythmic lines in simple, informative strokes, without looking at the paper. Try lifting one sheet of paper from the pad with your nondrawing hand, holding that sheet at

an angle that frees your drawing hand to draw on the next sheet under, but shields your eyes from your drawing hand. Partially covering your drawing page helps keep your eyes where they belong in this exercise, *on the model.*

Working on inexpensive newsprint should encourage exploratory impulses. Draw one sixty-second gesture after another with an eye for the structural flow of the model. Draw those sixty-second gestures, with the model changing his/her pose each time, for two hours, or the duration of a class period. Or, use any cooperative human being—a child, a spouse, a friend, whoever will pose. Be kind to your model. A five-minute break every twenty-five minutes helps everyone. And break time is the best time for reviewing your work.

**Figure 2.6** *Drawing the shapes of stress.*

**Figure 2.7** *A stress gesture showing stress in the right arm and leg.*

## BODY STRESS AND WEIGHT IN GESTURE

### Materials

*3B pencil or No. 2 conte crayon*
*Newsprint*
*A model*

To understand *stress gesture,* the student should assume the same poses that the model takes. Each of the poses should be ones with the body leaning, twisting, or bending more to one side than the other. A standard straight, up-and-down pose will not work for this exercise. As the model assumes a pose, so should you assume the same pose alongside your tables, desks, or easels. Hold the pose about thirty or forty seconds. Whenever a point of stress becomes pain, make a mental note not only where that pain

is, but also what the *shape* of it is. Having concentrated on that *pain's shape,* resume your drawing position, look at the model and try to incorporate the shape of the stress you just felt into the gestural information you are drawing from the model. For now, lights and darks should assume the role of *light meaning less* and *dark meaning more.* The heavier lines should indicate where there is more stress. Press down on your pencil to accent where you see *weight* of the body or *pull* of the muscles. The actual gesture will be somewhat more abstract than the former ones.

Drawing *stress gestures* does not mean drawing legs, arms, torso, and such. Learning to draw the figure has more to do with learning to see abstractions of the figure; in this case, drawing the shapes of the stress itself, the stretch or weight you perceive in the figure.

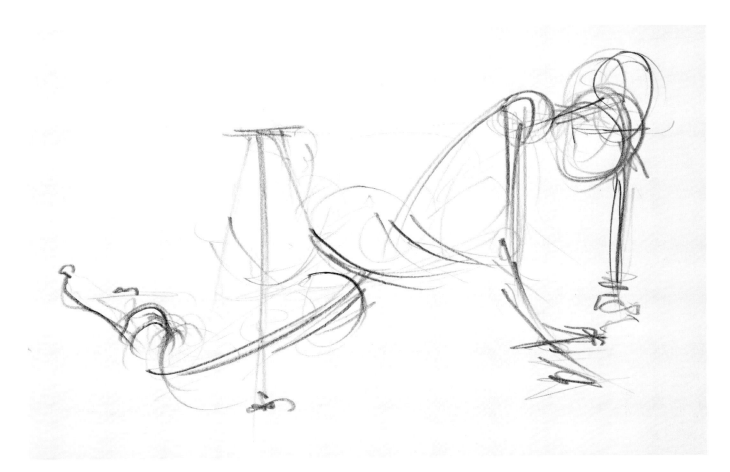

***Figure 2.8*** *A stress gesture showing weight and muscle stretch.*

**Figure 2.9** *Sighting vertically and horizontally while drawing a gesture.*

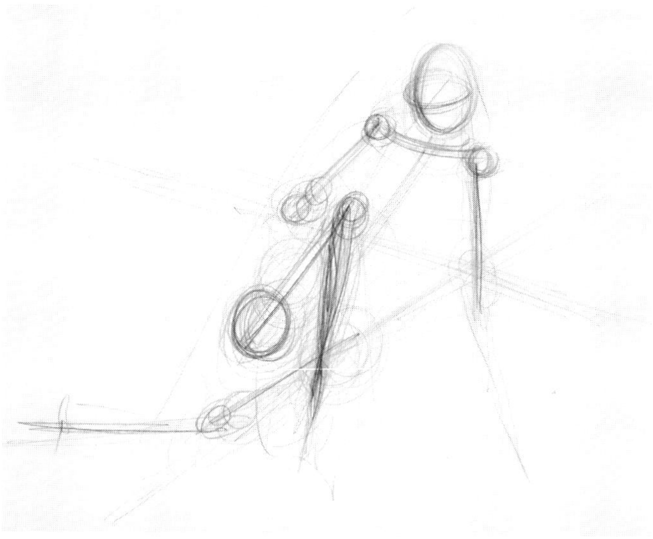

**Figure 2.10** *A gesture using vertical and horizontal sighting.*

## SIGHTING AND VOLUME GESTURES

### Materials

*2B or 3B pencil*
*18″ × 24″ newsprint pad*

The Blind gesture, the Stress/Weight gesture and now the Sighting/Volume gesture comprise the three gesture problems that focus the richest kind of information for a student to use when drawing either brief studies or sustained poses. This group of gestures teaches you to be both spontaneous and informed. A very real difference exists between a naive artist and an uninformed one. An uninformed artist simply does not know how to capture information. Building a solid base for retrieving artistic information will allow you also to control that information, to use and reuse it in other drawings, or to spin off into more sophisticated problems. Eventually you will develop your personal sense of space mastery.

When one *sights*, one imagines a graphing system of thin black tape strips making equal units like squares on a piece of clear plexiglass. The plexiglass window should be parallel to your eyes as you look at the model. That graphing system of vertical and horizontal lines is used to

locate points of the figure above and below each other and from one side of the body relative to the other side of the body. For instance, looking at the following example, you can see about where the elbow is to the knee, on a horizontal line. That sighting tells you that any part of the body above or below that line has to remain there. Now sight vertically, the same elbow to the knee of the other leg, the model's left leg. And, that sighting tells you that any part of the body left or right of that line has to remain there.

Remember that the shoulders are not joined at the neck nor are the legs joined at the spine. Early efforts often reveal that a student forgets how the shoulders connect to the upper sides of the body, and the legs connect to sockets in the pelvis.

If you look at the following example, you can see that this sighting gesture is begun with simple straight lines to indicate what angle a body part has to the floor. As you look at the model and draw those initial straight lines, let your eye continue vertically and horizontally, up and down and across the body. Let your pencil follow on the paper with your eyes on the model. Doing that helps promote a vitality in your line. Certainly, you may add sighting lines. These gestures are exploratory. Gradually, once the figure has been placed on your page, add volume. You can see

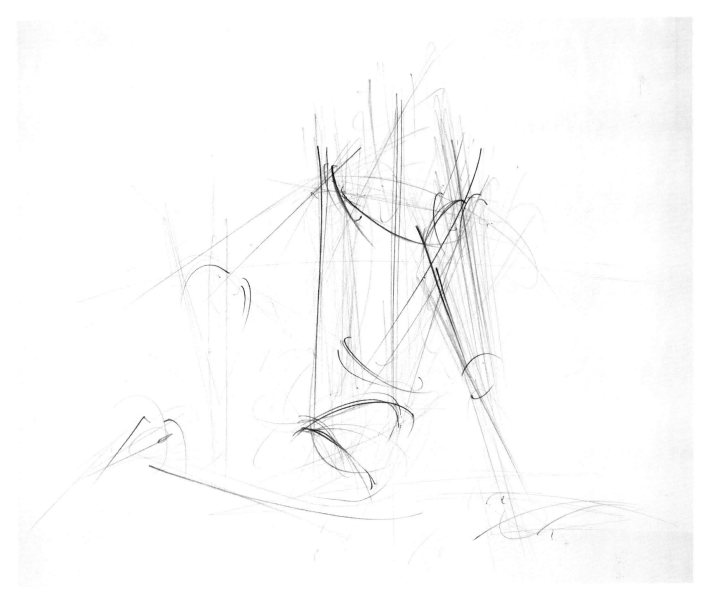

**Figure 2.11**  *Begin with straight lines, sighting lines, then volume.*

the beginnings of volumes in this example. Volume is drawn simply as though you could see through the body. The body has rounded masses.

Draw the volumes as if you could separate parts or sections from each other and see the section ends from where you are sitting. For instance, if you are looking at the figure and could separate a lower section of the calf from a higher portion of it, how would the ends of those sections appear from where you are sitting? Probably ellipses and not circles. Draw lines in the gesture to indicate those volumes. Have the model take a few poses and practice that sequence, straight lines, sighting lines, then the curving volumetric lines.

**Figure 2.12** *Begin with straight lines, sighting lines, then volume.*

Taking each issue separately—just lines, just sighting lines, just volumes—on separate pieces of paper, will help. Once you understand how to place straight lines on the paper at angles representing the relationships of body parts to the floor, then the sighting lines, then the volumes, and have done each of these individually, you are ready to put them all together in one drawing.

Begin, as you have, head to toe to toe, to hand to hand. Pencil still moving, sight for the angles and proportions. Keep your pencil floating over the whole body and try not to linger too long. Continue scanning, making markings where needed, to suggest the height of the knee, the placement of the hand and so on. Finally, finish off the gesture with indications of the mass, using lines that indicate geometric/volumetric masses.

As you can see in the example below, your choices teach you much about the volumes and placement of proportions of the figure. Try not to weigh yourself down with too much detail. Keep the poses for these problems not much longer than three minutes. The following example indicates how one of yours might look after assimilating all components into one gestural drawing.

**Figure 2.13** *Sighting and volumetric gesture.*

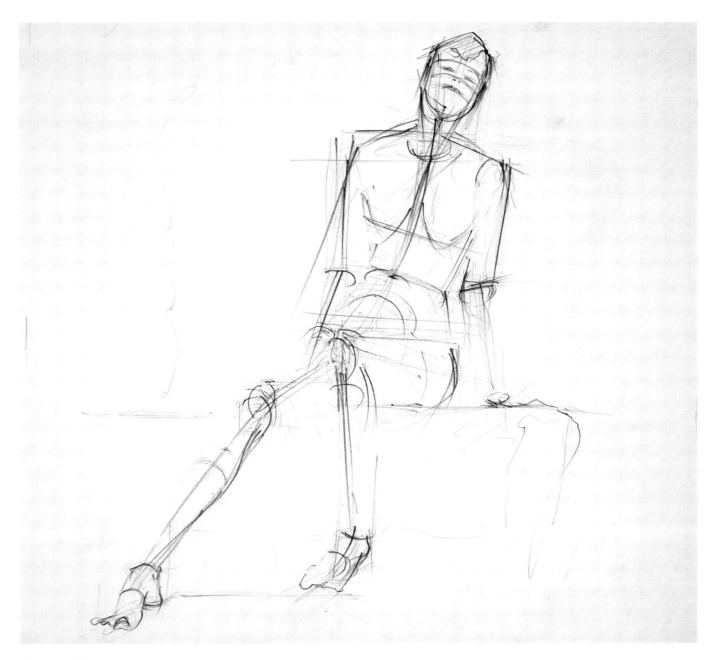

**_Figure 2.14_** *Sighting and volumetric gesture.*

**Figure 2.15**  *A gesture drawing too small for the page.*

## VARYING SCALE IN GESTURE

### Materials

*2B or 3B graphite pencil*
*18″ × 24″ newsprint*

As one moves through gestural concerns, varying scales become a problem. Drawing a figure well means using the size of the drawing sheet to accommodate the image of the figure before your eyes. Sometimes the image is too small. Sometimes the gesture fits.

If a model kneels with arms around the legs, that foreshortened figure on the page is larger in scale than the image of a model standing up, arms outstretched. Drawing is frustrating when we try our best to contain a figure on the page, only to have the legs, head, or elbows drift off because we did not plan things very well. Our best efforts come to nothing because, for some reason, we just could not keep the extremities in the space we had assigned for them. The solution lies simply in practice.

Have the model kneel in something like a fetal position. On your page make a mark for the top of the head. Make another mark for the base of the figure about one inch from the bottom. These marks set your scale. Now, make a mark for the farthest point of the body on the left

**Figure 2.16**  *A gesture that fits the page.*

of the paper and then, on the right as well. Lightly gesture the figure in, head to toe to toe to hand to hand. If your drawing stays small and in the middle, go through the process again. Once you get the first rhythmic proportional gestural lines down, you can go back into the figure for the angles, the sighting, the geometric shapes.

Now have the model stand with feet wide apart and arms above the head, stretched out. Try the same process again, head to hand to hand to toe to toe. You find, quickly, that the ratios of the figure must diminish in scale in order to stay within the boundaries you have framed. The head now is much smaller by comparison to the last drawing, the torso shorter, thinner, and so on.

Try this exercise over a period of ten minutes, alternating closed, fetal poses with poses that extend the extremities. Each pose should not last more than one to two minutes.

*Figure 2.17*  *Gesture scaled to fit the page.*

*Figure 2.18*  *Gestures showing accommodations for scale.*

*Figure 2.19*　*The pose.*

*Figure 2.21*　*What you try to see in your mind's eye.*

## REVERSE GESTURES

### Materials

*2B or 3B pencil*
*18″ × 24″ newsprint*

The *reverse gesture* exercises begin a series of types of reversals. This gesture exercise asks you to simply reverse the image. If the model poses with leg and arms extended up and left you draw him/her going the other direction. For these gestures, the model assumes a pose more to the right than left, or more to the left than right. Your drawing will appear in reverse of the pose.

Stretch your thinking to keep the pose accurate, to keep your eye sighting vertically from one point to another (nose to knee, for instance) and horizontally (shoulder to shoulder, for instance). Begin with the same head to toe to toe to hand to hand overall layout, then go back over the lines keeping your eyes focused on the model as much as you can, drawing the reverse body pose, putting in lines that suggest stress or weight.

*Figure 2.20*　*What your drawing should look like.*

**Figure 2.22**    *What you see.*

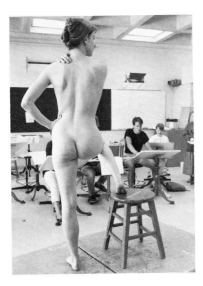

**Figure 2.23**    *How you should think.*

## GESTURES ON THE OPPOSITE SIDE

### Materials

*2B or 3B pencil*
*18″ × 24″ newsprint*

Begin with the model assuming a pose. As he or she holds that pose, draw the figure as though you were standing on the opposite side of the model, looking at his/her back just across from where you are actually working. Two things happen here. One, the pose will assume a reversal, but of a different sort than the reversal with which we just worked. Two, you need to think through what the opposite side looks like when the front becomes the back, or vice versa. Where are legs, arms, and such positioned

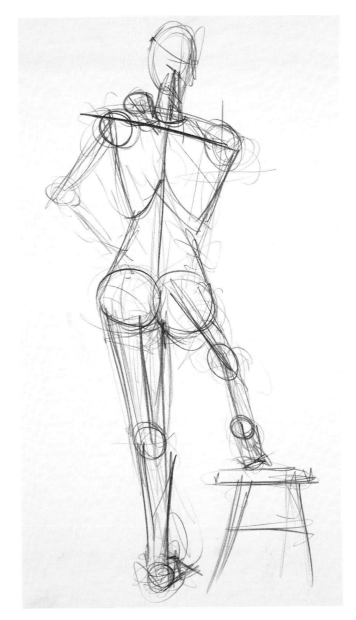

**Figure 2.24**    *The end result.*

now? Visual information is simply not given to your eye as it had been formerly.

Try a few. You will soon catch on. Begin with head to toe to toe to hand to hand. Place the angles, think through the volumes. Drawing the gesture, the angles, and the volumes are three of the richest processes for learning how to develop skills.

If you think this problem is an exercise in nonsense, let me suggest something. Many times artists need a *right hand holding a jar*, or *the inside of a left leg*, and so on as a reference for completing a work. Not having a model handy, and no pictures to help, they must rely on the "opposites" of what they do have. They may look at their own left hand, reverse it and draw the right hand in the drawing. The same for the leg. Experience suggests you do this gesture with regularity, and it will come to your rescue later.

## OTHER-HAND GESTURE

### Materials

*2B or 3B pencils*
*18″ × 24″ newsprint*

The tempting part of this gestural problem is to believe it will force the opposite side of your brain to operate, in contemporary lingo, to make a left brain/right brain transfer. By using your nondrawing hand, you do have to *think* more than otherwise in these exercises, simply because you have less control over line quality. By thinking more actively, you participate in ways you might not otherwise while drawing. While your finished pieces will appear to be more awkward from the lack of control, they also appear more spontaneous and more expressive, the result of that very thing, lack of control. Some of the finest student gestural drawings I have seen have been ones done by persons without their glasses, when they could not see well at all. Control is a funny thing. Sometimes it works, sometimes it doesn't. Knowing when to apply controls comes from ongoing learning experience.

For now, try again, the head to toe to toe, to hand to hand routine, then try to find the stress in the figure, pressing more where there is more weight or stretching, pressing less where you see and feel there is less.

Always bear in mind your composition. According to the poses the model takes, adjust your scale. If he or she is doubled into a fetal position, remember your figure will appear larger when you fill the page than if if the model assumes a vertical pose, arms overhead.

Drawing with your nondrawing hand should carry with it the same sense of scale and composition that the gestures have when you draw with your actual drawing hand. Your drawings may look similiar to the one pictured below, rather rough, uncontrolled, and spontaneous.

***Figure 2.25*** *Other-hand gesture.*

## TWO-HANDED AND REFLECTIONS

### Materials

*2B or 3B pencils*
*18″ × 24″ newsprint pad*

For this gesture the student needs one pencil in each hand. You will be drawing two images, simultaneously, connected to each other, like reflections. From your position relative to the model, decide whether your left or right hand will draw the model's actual pose, and which hand will draw its make-believe reflection. The reflection in your mind's eye can be to either side of the model, from the feet down as though a person was standing on a mirror,

or the head up as though the heads were attached. That choice will determine the place on the body where the "reflection" will meet the actual image. Look at the examples.

Some students split the figure in the middle, then draw a reflective image of one-half the body.

The best way to begin these kinds of gestures is at the "touching point" of the figure. Beginning there, move both pencils coincidentally as you continue to refer to the model. Since you have already experienced the awkwardness of using your nondrawing hand, this exercise should not surprise you by appearing less controlled yet perhaps more expressive. These gestures should be about two minutes in duration.

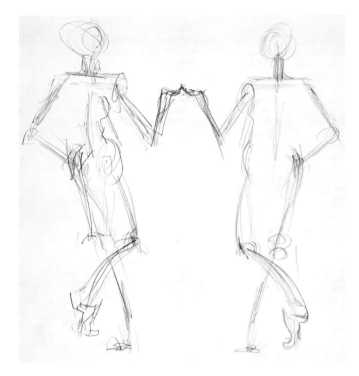

**Figure 2.26** *Drawing the model with left hand, reflection with right hand, side to side. .*

**Figure 2.28** *A gestural reflection, foot to foot.*

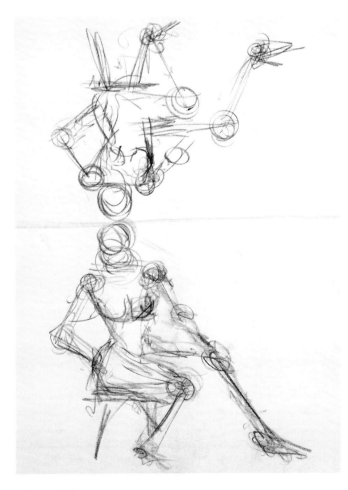

**Figure 2.27** *Drawing the model with right hand, reflection with left hand, top to top.*

**Figure 2.29** *A gestural reflection of one-half the body.*

***Figure 2.30*** *Action gesture of a figure, standing by, crouching on, and bending off a chair.*

## ACTION GESTURE

### Materials

*2B or 3B pencils*
*18″ × 24″ newspaper pad*

Our stroke-maturing purposes will be well served here by drawing a *three-in-one* pose, meaning that you will unify three poses into one drawing. The model will take three separate poses while keeping stationary one part of the anatomy, such as a hand, a foot, an elbow. The poses, for instance, can begin alongside a chair, the model standing

with the right foot on the chair seat. The model holds that pose for thirty seconds. The second pose might have the model crouching on the chair seat, still with the same foot on the seat of the chair. The model holds that pose for thirty seconds. The last pose could be the model bending over touching the floor with one foot and two hands, keeping the same foot in its position on the chair seat. The model holds that pose for thirty seconds.

As you draw, begin by placing your pencil where the stationary point has been identified and go from there for the first pose. Try keeping your eyes as much on the model as possible. The first strokes should be done quickly, head

***Figure 2.31***   *An action gesture of a figure sitting, leaning, and tilting off a chair.*

to toe to toe to hand to hand, just about as fast as you can say that to yourself. Those strokes set up the linear flow of the figure in gesture. With the next strokes you can use your pencil to adjust lines, gauge proportions somewhat, and again, vary the pressure of your pencil, more pressure for areas of stress, less pressure for areas with less stress. You have only thirty seconds for each pose, so the drawings should remain loose and unfinished. Keep your mind active with your hand. Try as much as you can to match the movement of your drawing hand with the movement of your eyes over the model. You can see in the examples that lines will overlap as you work one image on top of

another. This is what is supposed to happen. Overlapping lines and images suggest action. When the model assumes the second pose, repeat the process you initiated for the first pose, head to toe to toe to hand to hand, and continue. So, too, for the third and final pose. You can be selective in what you choose to draw, or you can emphasize, but your drawings should look something like the example shown.

Before you begin your drawing, have the model assume each of the three poses. That way, you can see whether you will need a horizontal or vertical format for the three-in-one image.

**Figure 2.32**   *A continuous motion gesture.*

## CONTINUOUS MOTION

### Materials

*3B or 4B pencil*
*18″ × 24″ newsprint pad*

The model for this gesture will need a model's stand or table that is approximately 6′ by 8′ for purposes of movement in slow motion. The model should think of the stand as providing above it a large cube of space that has height, breadth, and depth. Within that large three-dimensional space, the model should move continuously in extremely slow motion bearing in mind that the movement needs to extend behind, in front, to the sides, down to the feet, and up as high as the person can reach, in an ongoing variety of turns across the surface of the stand. The sequence is not unlike a free-form, very slow modern dance.

The student will need to turn his or her drawing pad because the vertical figure is now moving horizontally across the model stand, creating horizontal compositions.

As you draw the model in motion, try to see the whole figure, but concentrate on the major lines that indicate motion. Bear in mind that you will, again, not concentrate on one figure unit at a time, but allow the lines to lap and overlap, one into the other, as the case may be, to visualize the slow motion of the model.

Vary the pressure of the pencil when you see stress in the reaching or stretching of the model's body. More pressure, darker lines, less pressure, lighter lines. Feel with your eye, as you go, the mass of the body. Push lines up and around, over and down, all the while keeping your eye as much on the model as possible and as little on the paper.

When you come to the end of the drawing page, simply turn the page and continue onto the next clean sheet. Depending on how slowly the model moves, and how long he or she continues, you will have one to several sheets covered with drawings.

The end result will look something like the following samples.

***Figure 2.33***   *A continuous motion gesture.*

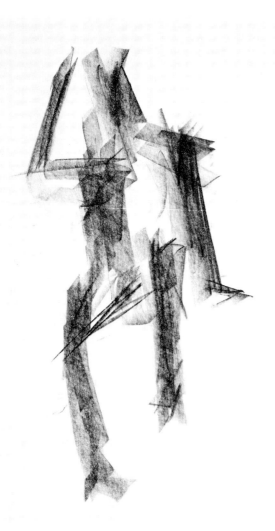

**Figure 2.34**   *Value gesture with awkward vertical lines.*

**Figure 2.35**   *Value gesture modulated with body forms.*

## VALUE GESTURES

### Materials

*A 2" piece of black chalk or charcoal placed on its side*
*18" × 24" newsprint pad*

Most of the gestures with which you have been working
have not included the use of value. The *stress/weight gesture*
suggested you apply more pressure on the pencil to indicate muscle stress or body weight. Now we turn to the use
of values to signify external and surface masses of the body.
You will be helped if the model will assume a pose in a
darkened room lit by a spotlight. If no spotlights are available, natural lighting from a partially closed window blind
should help indicate shadows and shading. These gestures
will take a little more time. The suggested time allowed
would be about three minutes for each page.

Once the model is settled, look at the person and squint
your eyes. Squinting adds contrast and helps sharpen the
sometimes subtle differences between lights and darks.

Notice the play of light over the whole body. Still squinting,
not yet drawing, seek out the darkest shadow shapes. Now
place the chalk or charcoal on its side so that your strokes
make wide lines. By making wide marks, you can more easily
feel through your eyes the distribution of the masses of the
body as you draw. In this gesture you are reminded that
the body is rounded. With the chalk or charcoal try to draw
the shadow as it lies across or around the physical shape
of the body. If you do not think of the shadow and shapes
*over* the body masses, you are likely to get a series of awkward vertical lines that indicate the edges of your chalk
rather than the play of lights and darks over a figure.

Try a few of these. Look at the model, squint your eyes,
place the chalk or charcoal lengthwise onto the paper,
moving the chalk or charcoal in such a way that you indicate body masses using dark shapes.

Now, to extend this exercise into something even more
beneficial for you, have the model take another pose within
the same lighting conditions, that is, a dark room with a
spotlight placed wherever you wish. As you look at the

**Figure 2.36** *Value gesture with figure in context.*

**Figure 2.37** *Value gesture with figure in context.*

model sitting, standing, or kneeling on the model stand, observe the shadows made by the body and cast against other surfaces. In your mind's eye as you squint, try coding like value shapes throughout the whole composition. Coding value shapes throughout a composition requires that you disengage from the objects and the model as physical units, as things in themselves to be drawn. Think of *just the shadows* before you as specific shapes. Each of those shadow shapes has a tone. Seek out all shadows of the same tone and draw only those shadow shapes over the whole sheet of paper, making a composition simply of *like* value shapes. No body, arms, legs, or drapery. Just *like* value shapes.

What this problem fosters is a maturing process in learning to abstract information while drawing the figure. The value shapes you see *on* the model are coded with any like value shapes that extend *from* the body, lie *among* the drapery or *in and around* any props. Look at *like* value patterns that appear to lie over or on anything. Once you have decided where the lightest value shapes are throughout the

set-up, draw them as shapes, *now using the mass* of the object to guide your strokes. For instance, if something is round, flat or bumpy, the value strokes should follow the surface feature of the object as the value shapes lie over it, on it, by it, or around it.

Look for the next darkest series of value shapes throughout the set-up and draw those same tonal shapes, *now using the mass* of the object to guide your strokes. Draw only this set of shapes throughout the whole composition, over your paper. Then, look for the next darkest value shapes, continuing the process described until you finish the exercise with the most dark of all the shadow shapes.

Do not draw the object and color it in. Coloring is not the point here. Look for like value shapes and draw them as they lie throughout the set-up.

By doing this, you keep the figure in its context and you create better compositions than those made with a single figure in the middle of the page. Another reason to work this way is to keep your mind on esthetic issues, lights and darks as a system, rather than seeing the body as a body with a head, arms, and legs. If you continually see the figure just as a figure, you will run into mind blocks early on that are difficult to overcome later. Trust the process, which is multi-layered. One step at a time brings you closer to understanding what your genuine capabilities are. These value gestures should run about three minutes.

## ELONGATED LINE GESTURES—HEADS

### Materials

*B pencil or very fine pen point*
*Smooth oaktag or similar type of paper*

One of the more interesting challenges in drawing pertains to heads. The heads we draw in gestural form do not have highly articulated features. When we begin drawing heads and faces, most of us want the picture to look like a portrait when we are done.

This gesture exercise is intended to reinforce masses, to work with volumes. Once a student can draw the head as a clear glass egg, and circle that egg shape with lines positioned to indicate eyebrow, eye, nose, ears, and mouth, completing the features is relatively easy.

The elongated gesture is a continuous gestural line wrapped around and around a form until the end result appears to be an image made out of a number of layers of thin thread. Sight the head to the neck. Is the neck vertical, the head tilted? Lay in the direction lines. Then place your pen point onto the paper, keep your eye as much on the model as possible, and begin to move the line as though you were wrapping a string around the head. Initiate a new line, a circular volumetric line indicating the brow. Draw one to suggest the base of the nose. And one for placement of the mouth. Then keep your pencil moving. Essentially you are feeling, with your mind's eye, the topographic undulations, the surface ups and downs, as you move your pen up and over and around both the front and back sides of the head. You are literally drawing around the model's head. Try to let your pen roam to different parts of the face and head. Just drawing circles around and around would make the end result look like an unimaginative slinky wire spring.

These poses should be about five minutes in length. The finished piece might look similar to the following examples.

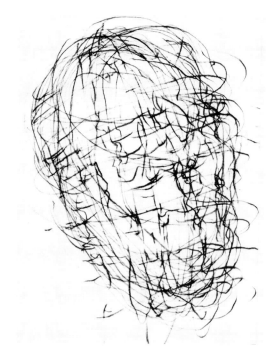

**Figure 2.38**   *Elongated line gesture of a head.*

**Figure 2.39**   *Elongated line gesture of several heads.*

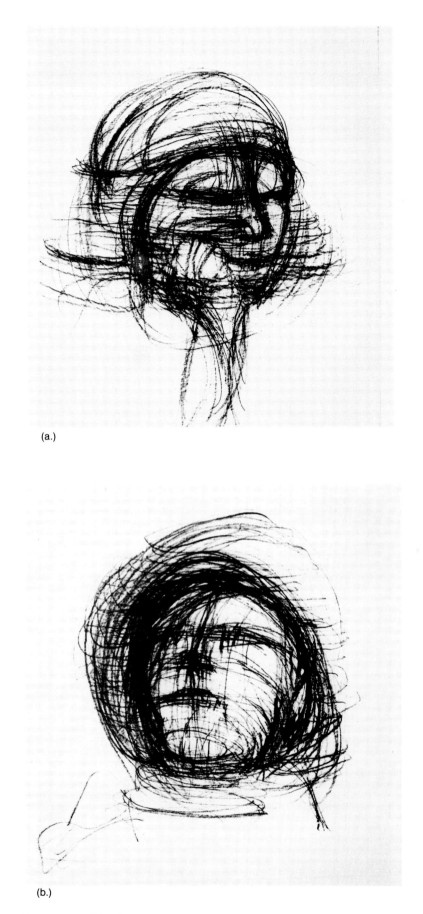

(a.)

(b.)

**Figure 2.40 a, b** *Elongated line gestures of heads.*

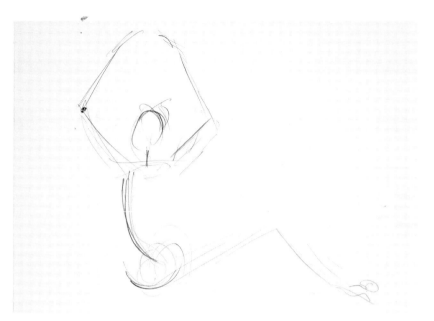

**Figure 2.41** *Memory gesture.*

**Figure 2.42** *Memory gesture corrected.*

## CREATIVE OPTIONS—MEMORY GESTURE

### Materials

*2B pencil*
*18″ × 24″ newsprint pad*

Memory gesture begins simply by looking at the model in a pose. Observe the model for about twenty seconds without referring to your paper. If you wish, in order to gauge the scale for your paper, you may want to trace a finger or thumb over your paper, using one or the other, or both, as a simulated drawing instrument. The finger or thumb is intended only to help you gauge scale and place-

ment when you continue. When you stop looking at the model, close your eyes and do not refer to the paper. Again, as in blind gesture, lift the bottom of one sheet of paper from the pad with your non-drawing hand, holding that sheet at an angle that will free your drawing hand to work on the next sheet under. Doing the gesture this way keeps you from sneaking a peek. You must rely on your memory.

Once complete, check your image with the model who has retained the pose throughout this time period and see how well you remembered ratios, proportions, placement/sightings, and such. Go back into the gesture and make your modifications, this time referring to the model.

**Figure 2.43**  *Two pencils in one hand.*

## CREATIVE OPTIONS—TWO PENCILS, ONE-HANDED GESTURES

### Materials

*Two colored pencils of your choice*
*18″ × 24″ newsprint pad*

Holding two pencils in one hand can act as one wider drawing tool allowing you to fill up more space in less time for this gestural variation. Drawing this way allows you to double your images, as you move across or around the paper. It can also make you look as though you have hiccups. Nonetheless, it is rather fun to try. With two pencils together, run the head to toe to toe to hand to hand, then the sighting, then the volumes. Thereafter, play with the image as you will, perhaps elongated, perhaps in value, perhaps combinations. Using two pencils in one hand is a wonderful way to work while the figure is in continuous motion, too. The model should be posing about two to three minutes, or longer, as your choices indicate.

**Figure 2.44** *Gesture with student changing positions, same pose.*

## COMPOSITIONS AS YOU GO

### Materials

*2B or 6B pencil or conte crayon*
*18″ × 24″ newsprint pad*

There are two sequences to this problem. Each will use one piece of paper. In the first sequence, the model will retain the pose while you walk around the model finding five different positions from which to draw, creating a composition as you go. During the second sequence, the model will change positions. You will too, but you will leave your own newsprint pad at your easel or stand, while moving to someone else's pad and drawing on theirs. By the fourth and fifth move, you will need to unify the compositions on the papers on which you draw. Varying your media might help. 2B to 6B pencils, an Ebony pencil, a colored pencil, and conte crayon are a few suggestions.

First sequence:

Have the model take a pose he or she can sustain for about ten minutes. You will change positions five times. Build your compositions as you go. Observe the model, draw a gesture, and move on. Go to another position, ob-serve and draw, to another position, observe and draw, and so on—all with the intention of building a composition.

The images you are building on the paper can overlap and indeed should overlap at some time. Otherwise you will find you have a relatively sterile composition with separate images sitting in a blank field of space. Remember that you can vary the weight of the line either by changing pencils or by placing more pressure for mass/weight/stress in the lines indicating those things your eye tells you are part of the model's stance. What about scale? Change that too, making some images large and some small. Turn your paper 90° four or more times and place images upside down or sideways to help build the composition. At the end of this exercise you should have a series of images that hold together relatively well as a unified work. The end result will probably linger close to the surface of the picture plane, meaning there will be little background or perspective incorporated. This exercise is sometimes difficult because one cannot always solve the compositional/spatial problem presented. But this exercise is given to help you see, select, and create options for yourself that you might use again later on.

**Figure 2.45**  *Gesture with model changing positions, student using different drawing pads.*

Second sequence:

Take your position at your easel or stand and leave your drawing pad there regardless where you move in the room. Each student will change to another student's drawing pad or paper and, as the model changes poses, so will the students change easels. The point of this drawing is to create something on a more spontaneous level, often trying by the fourth or fifth shift to tie the compositions together. This problem is not as easy as it sounds for there is great variety in perception, capability, and intention. This problem does allow you to share your skills and competencies so that others might be helped by seeing on their drawing pads what you have placed there.

Again, try changing your pencils for a variegated value in line. Try changing the scale, turn the pads, and *solve* the *compositional problems* as they arise—those running off the page, those with images that are too small, overlarge, too dark, too light, too big. If you work at this, you will find your own capability for solving visual problems growing. ▲

3

▼

# Rehearsing What is Known about Imaging

## A SHORTHAND PERSPECTIVE

**P**erspective is treated in this chapter as a simplified process, "shorthanded." For a more thorough discussion of linear perspective, one-point, two-point, and three-point, see the authors's *Drawing: A Studio Guide.* An abbreviated approach to perspective is presented here for those who have had either little perspective training or need to brush up on what they already know. We will start with some terms.

Most persons learning to draw use what is known as the Renaissance Window. The Renaissance Window, the *Picture Plane,* refers to a simulated piece of plexiglass that you hold parallel to your eyes and through which you look to see things like cities, roads, and people.

The boundaries of the picture plane, top/bottom and side/side, form your *Cone of Vision.* This cone defines the perimeters of what you can draw.

The center of your picture plane is your central and most concentrated area of focus called the *Central Sight Line.* The Picture Plane is *always* perpendicular to your Central Sight Line whether you look out, up, or down. The Picture Plane, the perimeter of the Cone of Vision, that is, top, sides, and bottom of the Picture Plane, and the Central Sight Line can be seen in this early Durer illustration of a shorthand perspective.

A time-honored shortcut to the use of a full scale perspective layout is the use of graphing lines taped onto the picture plane. With the Picture Plane held before you, perpendicular to your Sight Line, you can refer to the subject, the figure, through the graph lines. As you look, align certain areas of the graph to coincide with parts of the

**Figure 3.1** *Leonardo da Vinci (1452–1519). The Last Supper. Mural. c. 1495–98. Mural. Sta. Maria delle Grazie, Milan.*

figure. For instance, you might want to see where the arm joins the torso. You note that the vertical and horizontal graphing lines intersect a little to the right on the torso and above the armpit. On your drawing paper, which duplicates your Picture Plane, draw your own graph to correspond to the one on your plexiglass plane. Once you do that, simply make notation marks on the drawing paper that indicate the intersection points of figure and graph, thereby creating an image. Note Jocopo Pontormo's use of graphing lines below in his study of figures.

Chuck Close is a contemporary artist whose work is a continuous round of graphing to image. He uses photographs, color separations, and graphing scales to create huge images of persons, sometimes 10 feet × 10 feet. The example shows him at work on a watercolor self-portrait

**Figure 3.2** *Albrecht Durer (1471–1528).* A Man Drawing a Recumbant Woman. *Woodcut, 2.6″ × 7.2″ (6.5 × 18.3 cm.). Courtesy, Library of Congress.*

**Figure 3.3** *Jacopo Pontorno (1494–1556).* Study of Figures. *Black and red pencil blended on white paper, 8″ × 6 1/4″ (204 × 158 cm.). From Uffizi, Florence.*

**Figure 3.4** *Chuck Close working on "Self-Portrait/WaterColor." Photo by Geoffrey Clements. 1977.* Close Portraits *by Lisa Lyons and Martin Friedman. Walker Art Center, Minneapolis, MN. pg. 42.*

in 1977. You can see how he works his way down the painting (or drawing) on his left to create the much larger scaled work before him.

In the following examples you can see how Close set about his study of *Kent* in 1970. Note the graphing lines he used and how he laid out a series of colors below to begin his selection of the portrait.

However, photos are not used in life drawing class, generally. The better experience is to draw from a live model. One of his final studies of Kent is 100 inches by 90 inches. That's eight feet by seven and one-half feet for Kent's head.

When you are drawing from the model in your room, you will be codifying what you see into a graphing system and transferring information that is three-dimensional to an image that is two-dimensional. In doing that, remember that the Picture Plane is flat. The graphing marks are vertical and horizontal. Coordinating these lines on the picture plane with the graphing lines on your drawing, you can copy any image, near or far. You can also readily see

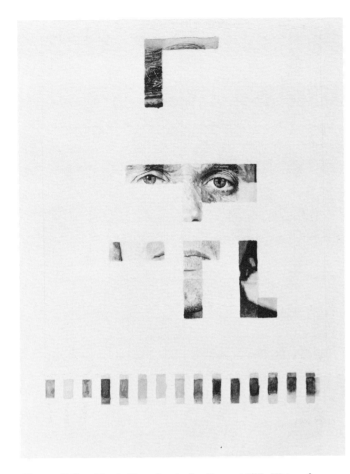

*Figure 3.5* *Chuck Close.* Study for Kent, *1970. Watercolor, graphite on paper. 30″ × 25 1/2″ (76.2 × 64.8 cm.). From Allen Memorial Art Museum, Oberlin College. Mrs. F. F. Prentiss Fund, 71.36.*

*Figure 3.6* *Charles (Chuck) Close. American, b. 1940.* Kent. 1970–71. *Acrylic on canvas (254 × 228.6 cm). Art Gallery of Ontario, Toronto. Purchase 1971. Copyright Chuck Close, c/o Pace Gallery, New York.*

where to place any lines that surround the model, such as table edges, chairs, or parts of the backdrop. The example below is of a seated model on a stand. If you look closely, you will notice that the table extends from part of the model's body. By superimposing a graphing system over your drawing, you can easily identify the angle of the table on which the model is standing. The more elaborate linear perspective modes for achieving the illusion of space may be short-circuited by the graphing process, with fairly accurate results.

Graphing does give you a limited control over the illusion of space. Utilizing it is simply a skill exercise. Mastery of this visual clue should help you, but you should not equate such mastery with criteria for what makes a work of art, a work of art.

In the gesture section you understood preliminary work with sighting alongside geometric constructs. Graphing is the same kind of work, probably with more elaboration.

If you use an actual plexiglass sheet, fine; many great artists have done so ahead of you. You are welcome to draw a graph on the sheet, or tape lines onto it. But using a literal picture plane is really not necessary.

First, study your position in the room relative to the model. Decide whether you want a picture plane close to or far from the figure. How much surrounding information will you include in this drawing? The table? A chair? Some additional material for draping? Once determined, you can easily lay in the gesture beginning with head to toe to toe, to hand to hand. Those lightly drawn lines will give you the top and bottom of your composition as well as the boundaries from side to side. You are setting your *scale*. Now you can begin your sighting, which means your graphing. For instance, align your pencil vertically with the tip of the nose and see what that line intersects below. A point on the chin? The sternum? A knee cap? A toe? If so, begin making marks that indicate those planes either with crosses or with short contour lines that infer the part of

**41**

**Figure 3.7**  *Model seated on a stand.*

**Figure 3.8**  *Overlay showing how to resolve drawing the table in a shorthand perspective.*

the body you have selected along that vertical graphing line. Continue sighting vertically and horizontally, making marks on the paper as you sight until you have finished. Once done, you might want to go back through the figure and put in very lightly the geometric shapes to help remind you of the specialized volumes of the body.

An early fault in drawing is forgetting that the figure has mass. Because students work on a flat plane, too often their figure drawings tend to look flat. Thinking in volumes continuously will promote good drawing habits. Sighting is essential. So is thinking in volumes. Once you draw in volumes, you can flesh out the figure. Continue

looking at the model as much as possible. The information is there.

Forms on the body, because the body is muscular and soft, change when pressure is applied, or if something on the body hangs upside down, or if the model lies on a feather mattress and sinks. As you begin to "flesh out the drawing," note the contours of the superficial muscles (the surface muscles) and draw the contour of the body as well as you can.

Following are three examples showing sighting, volumes, and the fleshed-out figure.

**Figure 3.9** *A figure with sighting lines and volumes indicated.*

**Figure 3.10** *A figure with sighting lines and fleshed out volumes.*

**Figure 3.11** *A figure with sighting lines, volumes, and fleshed out forms.*

# COMPOSING

Composing a picture has to do with ordering, which means choosing. You select and arrange materials, artistic elements, strokes on the support (such as paper or canvas), and the subject matter. The finished art work in a traditional sense has a composition. Let us look at some of those components.

*Materials* are canvas, paper, wood, the "something" on which you ready a surface to draw. What is used to draw with are *media* such as chalk, ink, or charcoal. The marks or strokes the artist uses to create picture *space* incorporate the formal elements, which are *line, value, color, texture,* and *shapes.* The *subject matter,* that is, still-life, landscape, figure, or abstraction, is selected in whole or parts and adjusted in the composition accordingly.

The compositional issues we will discuss here are related to the Renaissance Picture Plane, the illusion of the window. In the twentieth century some art theories branch to other spatial concerns, like "field painting," among which could be listed artists like Jackson Pollock. But Renaissance composition, in its most standard sense, used cross sections of lines that divided a sheet into thirds horizontally and vertically.

The four "crossings" indicate areas where a center of interest for the composition could be placed. The surrounding space may then be ordered in such a way that the viewer's eye moves to the center of interest.

By overlaying lines on top of Wesselman's work, a contemporary piece using Renaissance ideas, you can see how this simple nine-sector guideline is used to advantage. The upper left crossing line rests at about the model's left eye. The center of the composition is strong, in part because the eye of the figure is looking out at us. Let us analyze this composition using the formal elements:

1. Line—the diagonal line of the head intercepts an opposite diagonal line of the eyes. A horizontal line from the right converges with another diagonal line from the top center, apparently the outline of a lamp shade. The horizontal edge of the fishbowl moves the viewer to a background diagonal shape, taking the eye to the lamp shade, then to the woman's eye. The fish are horizontal lines, each with eyes to the left, moving the viewer to the diagonal of the throat and head and up to the eyes. On the lower left is a flower incorporating another diagonal line that points to the face. The lines of the rose in the upper left area are complex and rounded in a place where there are few

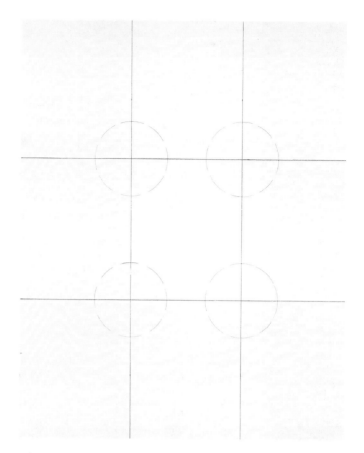

***Figure 3.12*** *Vertical and horizontal divisions as centers of interest for composition.*

other focusing lines. But the rose also draws attention to the eye. Finally, the hairline, which frames the face, is repeated in the curved lines of the eyes.

2. Value—The lightest values that surround the face all point to the face, the lamp shade, the horizontal edge of the fish bowl, the fish, the diagonal stick in the water, the floating tendrils, the stem of the rose on the bottom left, and the background horizontal value shapes. The darkest values set off the lightest values and repeat the harmony of direction. The background curtain shape, the shape of the hair, the throat ribbon, the two dark nostril shapes repeated in the two much larger irises, the downward turning eyebrows, all because of the dramatic contrast in values, darks against lights, maintain a cyclical rotation around the center of interest, the eyes, and most especially, the eye to our right (the model's left eye).

**Figure 3.13** *Tom Wesselman (1931– ). Bedroom Painting #43. 1979 Courtesy Sidney Janis/VAGA.*

3. Color—Color is not describable here since our image is in black and white. If it were in color, we could speak about density, intensity, and hue, all of which help center interest on the eyes.

4. Texture—Texture in this work is most dramatically revealed in the eyebrows. The slightest separations of the strands of dark hair against light skin reinforces the center of interest. Nearly all other textures have been removed, flattened, or barely suggested through lights and darks, as in the upper rose. Because of the understatement of surface textures, when texture is used it becomes rather dramatic.

5. Shapes—Shapes in this painting are largely flat and remain close to the Picture Plane. The face is as flat as the lamp shade. Only with the barest suggestion does one see modeling, or value used to suggest mass. There is so much activity at the surface that one struggles to understand how the volumes of the bowl, the bottom flower, the model's body, the lamp, and the curtain can all be included in such compressed space. Only with effort can we move past the shapes, the lines, and the values to pause ever so briefly in the background beyond the curtain.

# SOME SPATIAL OPTIONS

The following illustrations are examples of ways to use space by working with the Renaissance Picture Plane. These examples are intended to help you see many ways to work with space. Most of what you practice in these early stages are skills for looking at the figure, drawing quick gestural studies, and achieving perspective. Now that we have come this far, the idea of selective uses of space may help you in work you will do later on. And it may give you ideas of your own. The illustrations are not an exhaustive sampling, but simply a limited and brief exposure to some types of space used by artists, most of whom are contemporary.

*Ideal* is a term whose origins have Greek roots. In art the term suggests a selective process by which the artist works from nature (the most beautiful human models, for instance) and then improves their bodies and features. Among artists who have used selective methods for finding *the most beautiful body* are Zeuxis (latter 5th century B.C.), Alberti (1404–72), and Durer (1471–1528). Along with the selection by artists of beautiful males and females, often the artist incorporated mathematical systems that defined bodies with lines, circles, and such, creating *perfect* figures. Classical artists were trying to find perfection through mathematics and harmony.

Albrecht Durer's engraving, *Adam and Eve*, 1504, is such a work and is shown below.

Durer constructed these figures with arcs and circles . . . "lines and pinholes of the compass can still be discerned on the drawing."[1]

***Figure 3.14*** *Albrecht Durer (German, 1471–1528). Meder 1. Adam and Eve. 1504 Engraving. (trimmed to plate mark) .249 × .193. (9¹³⁄₁₆ × 7⅝) National Gallery of Art, Washington. Gift of R. Horace Gallatin.*

***Figure 3.15*** *Martha Mayer Erlebacher.* Adam and Eve. *1975. Oil on canvas. 64″H × 40″w. Sloan Fund Purchase. Sloan Collection of American Painting, Valpariso University Museum of Art.*

***Figure 3.16*** *Jack Beal (American, 1931– ). Still Life Painter. 1978–79. Oil on canvas 49¾ × 60 in (126.3 × 152.5 cm). The Toledo Museum of Art; purchased with funds from the Libbey Endowment, Gift of Edward Drummond Libbey.*

Our other example of idealism shows a contemporary spoof of Albrecht Durer's *Adam and Eve.*

While Erlebacher works tongue-and-cheek with her figures thematically (notice Adam and Eve pull apart rather than come together), she uses a refined nature as her background space. The figures are close to the Picture Plane, they appear relatively static, they are well-proportioned, and they have smooth musculature.

*Realism* has many meanings. The term can be used as the opposite of anything abstract, distorted, stylized, or idealized. For our purposes, realism means individualized, illusionistic as in the use of perspective (both linear and atmospheric), and something suggesting everyday life in some way.

Jack Beal's work is just such a piece. The receding lines of the table indicate linear perspective. Placement of vines, fruit, basket, and figure seem to follow a natural order of spacing. If you look closely at the example, you will see a girl peering through a group of leaves. She is so much a part of the still-life that seeing her is difficult simply because chair, plant, fruit, basket, table, and quilt covering are in front of her. The work does suggest everyday life physically as well as psychologically. The picture could be a metaphor for a cluttered life, a symbolic way for hiding ourselves from others. The picture might be a realistic portrait painting. The image is representational and the picture plane includes the strong use of the Renaissance window.

Another form of Realism is the use of *Atmospheric Perspective.* Most often, atmospheric perspective (which is related to linear perspective in that space appears to recede) is just that, atmosphere; (for example, outdoor scenes of some distance.) Colors and textures diminish or disappear altogether as the figure appears smaller and further away. Often, as in Caspar David Friedrich's *Monk by the Sea*, when a figure is used, it is small and nature looms large.

**Figure 3.17** *Caspar David Friedrich.* Monk by the Sea. *Oil on canvas. From National Galerie, Berlin. 1809–10.*

A third variation of linear perspective, or Realism, is the use of *Overlap.* This is very much like the system devised for you to use on the vertical/horizontal scale for sighting. An obvious use of overlapping is seen in Sandy Skoglund's *Life After Death (from True Fiction).* To the far right is an abrupt and enlarged nose and mouth over a woman's face. A hand appears in the middle of the picture, which looks as though it is actually resting on the picture plane, palm facing us. The hand's placement overlaps the car. The car is in front of the fence, and so on.

*Surrealism* sought to use another form of Realism, that of the subconscious and the dream. Surrealistic artists were on the frontiers of expanding the matter-of-fact, the physical presentation. They wanted to move away from logic to impulse with no moral or aesthetic boundary. In Mark Greenwold's *A Family Tragedy,* one sees a left-sided one-point perspective with a male and female inside a room.

**Figure 3.18** *Sandy Skoglund.* Life After Death (from True Fiction). *Acrylic on canvas, 48″ × 72″. From Leo Castelli Gallery. 1986.*

So far, the image is realistic. But the male and female are both too large for the interior. A theatrical tragedy seems about to take place, and a strange woman's decapitated head is floating as though alive into a ceiling well. Nothing makes sense. The expressions on the faces invite humor, though the situation presented is not really humorous. The doll's house architecture is strangely accommodating.

In *Equivocal Space,* which is what we see in M. C. Escher's *Drawing Hands,* the artist's left hand is drawing his right drawing his left. Equivocal space is space that is capable of more than one interpretation, it is ambiguous and inconclusive. Descriptive words attach themselves to equivocal like indeterminate, undecided, ambivalent, questionable, and dubious. Escher is the foremost proponent of this kind of space for image making. Many of his works indicate the illusion of deep but odd space. There are several good ways to work with Equivocal space, which we will discuss later. In this section we are simply illustrating possibilities.

**Figure 3.19** *Mark Greenwold.* A Family Tragedy. *Gouache on paper, 6 3/8″ × 7 1/2″. Courtesy of Phyllis Kind Galleries, Chicago and New York. 1984–85.*

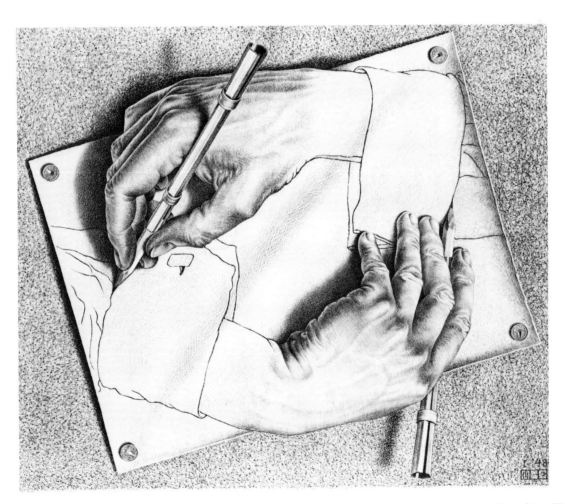

**Figure 3.20** *M.C. Escher (1898–1972).* Drawing Hands. *1948. Lithograph, 11 3/4″ × 13 3/8″. National Gallery of Art, Washington. Cornelius Van S. Roosevelt Collection. B 26939.*

*Contradictory Space* is space that uses the illusion of three dimensions, then flattens the same space one way or another. Either the artist will use large flat areas of color, will not change the scale of figures, or will use shapes indicating recession, then contradict the recession by making everything appear to stand up. For instance, our normal impression of a river is that it is blue or green with a shore line that moves into the distance. If we painted that water a bright red, from near to far, the shape of the river would appear to stand straight up and down on the canvas. Such is the space developed in Alejandro Colunga's work, *Nino loco tocando un concierto nocturno.* The piano is painted with a recessional line, the only object that appears to recede. The rest of the work is done as though everything in it were vertical or horizontal. Some use of overlapping indicates depth, but illusion of recession is occasional and contradicted.

*Simultaneous Space* is a reference for Cubism. The Cubist originators, Braque and Picasso, drew from nature, like the Greeks, but drew facets of the figure from all points of view, front, back, side, top, and bottom. The emerging image appeared fractured. Suddenly, a secure, singular picture plane was broken into a million pieces. No longer would artists need to contain images in the static presentation of a window. No longer would they be bound by a prescribed naturalistic, idealistic, or realistic boundary. Space was theirs to be had.

Russell Connor's work, *The Kidnapping of Modern Art by New Yorkers,* is a capricious work, a variation of Reuben's *Rape of the Sabine Women,* which is truly a monument to the picture window. Connor's men dash in and carry off Cubist mademoiselles from Avignon. You remember that Cubism destroyed the Renaissance Window. The figures here in metaphor humorously, but accurately, imply that New York finally has captured, and has become, the center of modern art.

**Figure 3.21**   *Alejandro Colunga.* Nino loco tocando un concierto nocturno. *Oil on canvas, 63″ × 79″ Private collection, Monterey, Mexico. 1984.*

**Figure 3.22** *Pablo Picasso (1881–1973).* Les Demoiselles d'Avignon. *Paris (begun May, reworked July, 1907) Oil on canvas. 8′ × 7′8″ Collection, The Museum of Modern Art, New York. 1906–07. Acquired through the Lillie P. Bliss Bequest.*

**Figure 3.23** *Russell Connor.* The Kidnapping of Modern Art by the New Yorkers. *Oil on canvas. 1984–85.*

## SCALE

The scale of a work is a surprising phenomenon. Try your imagination here for a little while. If you draw an image of a pair of eyes on paper the size of a stamp, your eye sees the pair, but without using a magnifying glass rarely makes out the strokes that create the image. Because of the smallness of the image, you probably would not be affected much, one way or another. Even if the eyes were a direct gaze, the image could easily be put aside. But, if you draw those eyes life size, their impact would be stronger psychologically, especially with a direct gaze. Some paranoia might be in order. Once we moved beyond our own response, we certainly could see the artist's techniques and materials and marvel at the reconstruction or realistic presentation. The image tends to have a strong impact.

Now, if you drew those eyes so that the distance between the outer edges was four feet and placed them at your own eye level on a wall in your room the impact would be quite significant, and you might be asked, at least by visitors, to place a covering over those two things piercing them with their intense gaze.

Making each eye larger still, say ten feet across, shifts the drawn eyes into abstractions. They become less obtrusive simply because a viewer cannot take in that large an image at close range. Putting ten-foot eyes on a billboard removes the intensity felt at close range and makes the scale work at a distance. But in a car driving by, you are less likely to be affected by the image, largely because billboards assume the principle of the quick glance. Fine Art, on the other hand, wants you to pause, to think, and to feel.

John Weinkein's miniature of an American Indian invites us into an intimate space. We are caught by the diminutive size. Most images we view are larger than this one. We are impressed that he did so well with such a small amount of space.

Chuck Close, on the other hand, works often in a super life-size scale. The example shows him in his studio alongside two portraits of friends. The works are about eight feet high by seven feet wide. If you see one of these paintings in a gallery or museum, you have difficulty in getting "past" it.

Because the portraits are frontal, the gaze of the eyes follows you wherever you go. The intense colors, along with the huge head, the eyes, and their gaze cast a spell, for better or ill, and moving past or disregarding the image is nearly impossible. Close himself says this, "I want to build a powerful image without using powerful gestures."[2]

***Figure 3.24*** *John Weinkein.* American Indian. *Drawing. 1988. 1 1/2″ × 1″.*

Students need to explore possibilities with scale. Staying with an 18″ × 24″ format can become dull. Using a life-size scale is challenging. There are some things to remember, however. One is finding paper large enough. Often students have used white bed sheets stretched over stretcher bars, like a canvas, for paintings. Their media ranges from ink to charcoal to conte crayon. The stretchers are usually leaned against a wall to provide an improvised easel.

The model needs to assume a pose fitting the shape of the drawing frame. Usually that means a seated, foreshortened position. If the figure is too high or long and drawn in life-size scale, the person drawing at close range has to move up and down the figure or follow the model head to toe. This moves your eye level, the cone of vision, etc. Keep the figure unit compressed to some degree.

Many students have used mirrors to reflect their own image for a self-portrait, making very interesting divisions of space. Our example shows a woman seated in the corner of a room.▲

## ENDNOTES

1. Walter L. Strauss, *The Complete Drawings of Albrecht Durer,* Vol. 2, Abaris Books, New York, 1974, pg. 756.

2. Lisa Lyons and Martin Friedman, *Close Portraits,* Walker Art Center, Minneapolis, 1980, pg. 16.

**Figure 3.25**   *Close in his studio with* Joe, *1969 (left) and* Bob, *1970 (right). Photograph by Wayne Hollingworth.* Close Portraits *by L. Lyons & M. Friedman, pg. 61.*

**Figure 3.26**   *Life-size scale drawing. 5' × 4'.*

# 4

▼

# Line/Shape

## BLIND CONTOUR/LINE WITH VALUE

### Materials

*2B or 4B pencil*
*18" × 24" oaktag or smooth paper*
*A model in a set-up*
*Time: About 30 minutes for the contour*
*Several hours for finished drawing with values.*

We are at a stage now where the exercises in these pages become more elaborate, longer, and generally more satisfying as your skills grow.

The *Blind Contour* is just what its name implies. We ease ourselves into working with a single contour line by observing and by concentrating only on what appears to be the outer edge of the figure. Obviously, human beings do not have edges, but we draw what appears to be an edge. A line drawn with greater pressure will be darker than a line drawn with less pressure. Lines, lights and darks, should suggest what weight or mass is behind the *edge*. If your lines do not carry that sensitivity, the figures created could appear as a flat paper-doll cutout, vertical and horizontal.

Since this is blind contour using a 2B or 4B pencil, you will need to use a relatively smooth paper surface, like oaktag, on which to draw. The reason for a smooth surface is assurance that the line will remain intact. Often on rougher papers the graphite will splinter into the fibers of the paper leaving the line less informative and appealing.

Once the model is posed, seek out your composition. Is the set-up more horizontal then vertical, or more vertical than horizontal? Adjust your paper accordingly. Again, you need to hold a single sheet of paper above the paper on which you draw. Drawing *blind* keeps your eyes on the model all the time, with very little control over your drawing, which is what is important in this exercise. Most students have a difficult time not looking down, referring to their work. But drawing *blind* means that you do not look at your work at all.

The point of this exercise is to sensitize your eye to the information carried by the variety of exposed surfaces throughout the body. Hair, for instance, has a much different texture than fingernails. Skin is dissimilar to the eyeball and so on.

Find a point within the composition before you, then place your pencil on the paper and begin. Once you begin, DO NOT refer to your paper at all. Feel free to lift your pencil when you need to, but DO NOT refer to your paper. The object is *not* to render a photographic copy of the set-up. The object is to *sensitize your eye* to surfaces and what they show you. You need not be worried about end results looking like something. That simply is not important here.

The more important and concentrated effort should be the varied qualities in your line. Does your line indicate what is being drawn? Is what you are drawing soft, hard, straight, curved, squashed flesh, or taut muscles? Be sure you incorporate the background lines of the set, too.

Once you have finished drawing, look at your work. If you have gone about this problem as you should, it will look strange, assuming you had the willpower not to peek at the work and try to *make it look right*. Just see what you have done by the way of sensitive lines. Did you vary the pencil line by pressing harder or less hard according to parts of the body receding or protruding respectively? The shapes you have on your paper may be comical, with legs appearing like pencils or heads like balloons. The figure images will be distorted but usually do retain the appearance of a figure.

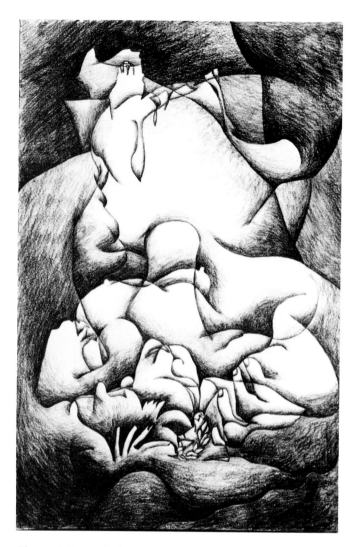

*Figure 4.1*  *A blind contour with value.*

*Figure 4.2*  *A blind contour with value.*

If your composition is more to one side of the page than the other, see if cropping some of the paper will help. Sometimes cropping can salvage an otherwise lopsided composition. However, if you cannot, do not worry. Find another point of view and begin a new drawing, taking care again that you concentrate wholeheartedly on effecting a sensitive line, which means, of course, that you draw slowly and you do not refer to the paper at any time.

You would do well to complete one drawing every twenty-five minutes. By the end of a class session, four to six of these drawings could be done. If you have less, you probably have done a wonderful job of not looking.

Bear in mind that you may ask the model either to keep the same pose or change it as your time structure allows.

The next part of this problem is the more relaxed portion.

Value, of course, is the gradation of tones from lights to darks. As you look at your completed drawing, ask what you want to do with the values. Do you want to have the strange image colored in with natural lighting from one side or the other? Would you rather infer a luminous quality making all edges appear dark, the centers of those shapes light? Or, the reverse of that, the edges light, the centers dark? Do you want the whole composition to begin with a dark edge and move to a light center? Or you could flatten the spaces into pattern shapes of light and dark. Try whatever solution seems to work.

In the following student works, you can see how drawing blind brings about a new vocabulary of images. You can also see what lighting effects the individuals chose to finish the pieces. All three are very well done. Each student has his or her special stamp on the work because no one else can draw the way they do, nor the way you do. Your choices make your style, eventually. This book is written to help guide you to expressiveness. Right now you are in the process.

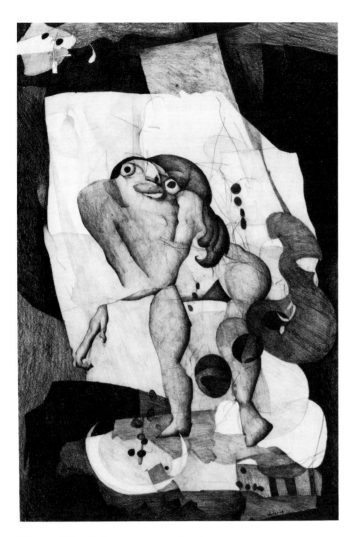

**Figure 4.3**  *A blind contour with value.*

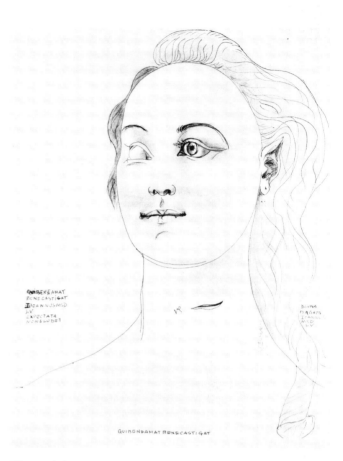

**Figure 4.4**  *John D. Graham.* Quibeneamat Benecastigat. *Ink and pencil on gray paper. 16 3/8″ × 13 3/8″. (41.6 × 33.9 cm) 1955. The Phillips Collection, Washington, D.C. Signed and dated. Gift of the artist, 1955. No. 0830.*

Try your hand at the finishing processes. See what unique lighting combinations you can use. A suggestion here: should you have any lines that do not define shapes, that is, if the lines are just left floating, try to draw them to a closure. To work with values in this problem you need to have a specific and enclosed shape. Resolving lines left dangling by closing them to become shapes will help you realize a better work.

While the following examples are not blind contour, they do show contour with the use of value. John D. Graham's strabismic woman (meaning vision disorder), with her seemingly enlarged body, appears to come from the timeless and monumental themes of the Renaissance, car-

rying aspects of the mythological. The picture's oddity, apart from the eye and the implied scale of the figure, rests in the very limited use of values. Only those of the eye and mouth have been brought to completion.

David Remfry's *Jubilee Rose* incorporates a beautiful contour line, which leads the eye up to the tousled hair and colorful red rose. The use of value is stark and dramatic.

Leonard Baskin's contour is much coarser in his work, *Watching Man*, in keeping with the coarseness of the figure portrayed. Value, though limited, plays with textures in the face, beard, and crotch. Similar, rather harsh lines of value cover the legs. An ink wash plays over the background, allowing the figure to stand as a silhouette.

**Figure 4.5** *David Remfry.* Jubilee Rose. *1978. Mercury Gallery, London, England. Published by Artist's Cards, London, No. 69.*

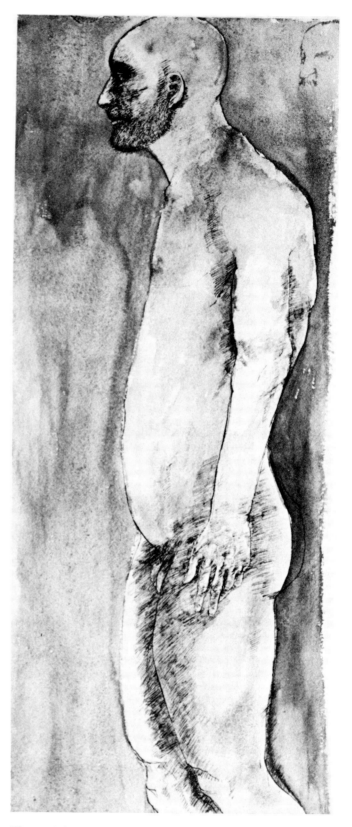

**Figure 4.6** *Leonard Baskin.* Watching Man. *1975. Ink and wash, 78.7 × 59.7 cm. From Kennedy Gallery, N.Y.* Naked to Nude *by George Eisler.*

# FIGURE TURNING

## Materials

*2B or 3B pencil*
*18" × 24" white drawing paper*
*Model posing in vertical standing positions.*

There are many images throughout history that incorporate time by making figures appear to move. The moving picture, still film images that move rapidly, is the obvious extension of this art form today. However, in drawing, we can create a sense of someone moving by using repetitions of a turning figure across a sheet of paper. The gestural interpretation is seen below.

If we use a contour line, we can slow the sense of motion almost as though we were taking still photographic shots and arranging them successively so that the figure appears to be in units of stop action. The gestures let us feel the momentary movements of the model turning.

Using language to describe this upcoming process may leave you somewhat confused. If you refer to the following examples, perhaps that will help clarify what the model needs to do.

The model should take eight poses that will complete a circle. Each pose is exactly the same as the last, except that he or she turns 1/8 of a circle each time. You can mark the tabletop with tape using a compass or ruler, dividing the circle into fourths, then eighths. After ten minutes the model shifts his/her feet to align them with the line of tape marking the next division of the circle. The model moves again after ten minutes, realigning his/her feet, keeping the same pose. The model continues this until a complete circle is made.

The student needs to see only part of the body. Pretend for a moment. If you close a door within five inches of the door frame, then ask the model to step into that five-inch area, you would see only part of the body. You would draw that. Ask the model to turn, replacing her feet, on

***Figure 4.7*** *Gesture of a turning figure.*

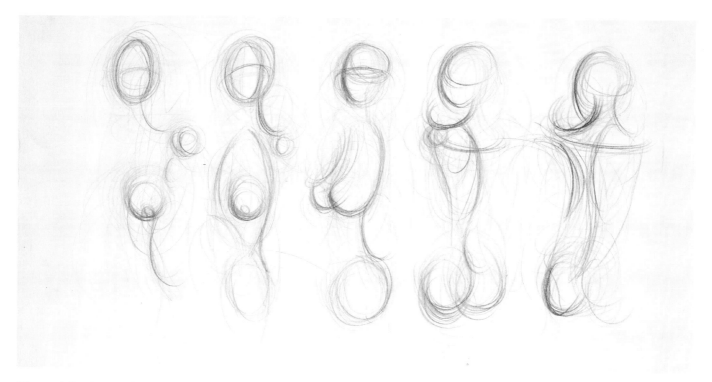

**Figure 4.8**   *Gesture of a turning figure.*

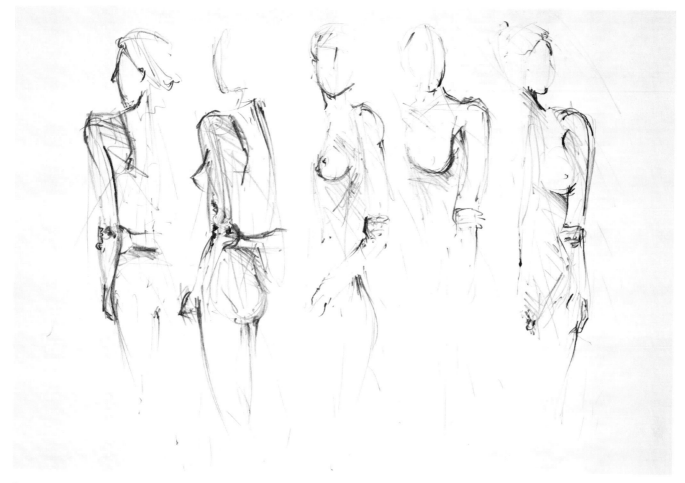

**Figure 4.9**   *Turning figure in space.*

the next taped strip of the circle as in the original pose. The model's body section is changed in that five-inch *open-door* area. Draw that part of the body just to the right of the first drawing on the paper. Continue to ask the model to shift poses in the marked circle while you keep your five-inch *door open* to the next portion of the model's body showing in the doorway, until you have completed the drawing sequence.

For instance, you have chosen to look to the right side of the body. You see the back of the model's right arm. You draw, as stated above, the shoulder, arm, right side of the rib, the waist, hip, and leg. The model turns. Now you see through the *opening in the door*, the model's scapula, back rib, waist, buttocks, back thigh, calf, and heel. The model turns again. Now you see the back of the head, the spine, the waist and buttocks, and so forth.

The object of the exercise is to use consistent vertical portions of a figure as it turns in space to infer on the paper that the image appears to be turning.

One concern students have with this problem is what to do with the negative space created between the portions of the vertical image. Some leave it white, others fill the background in with pencil strokes and others use classroom space as a background.

A word of caution here. Remember that as you draw each image, you need to keep head, shoulder, waist, knee, and foot at about the same place as before. If you are not mindful of a consistent scale, you soon might find that your heads are enlarging, the torso is lengthening, the thighs diminishing and your efforts will come to an odd end. So, as you move across the horizontal sheet of paper, remember to retain a consistent scale.

There are variations with this exercise, as there are on almost any of the exercises in this book. Placement of the image can vary on the paper moving vertically in a spiral. You can purposely change scale. You can select alternate sides of the model. The examples below suggests some variations.

***Figure 4.10*** *Variations for model turning.*

# LOOKING FOR ANGLES/FINDING INTERSECTIONS

## Materials

*2B or 3B pencil*
*18″ × 24″ white drawing paper*
*Model in a pose where arms and legs cross over sections of the body.*

Finding intersections is another way to help you sight, to help you retain a consistent scale, and to keep your eye moving from one side to the other as well as top to bottom. Without being able to understand and use proportioning, often the figurative image appears uninformed. There are, of course, thousands of drawings with distorted figurative images. But in the world of great fine art, those images carry meaning. An uninformed drawing is nothing more than that. Distortion is open to all of us who want to express a particular aesthetic issue, but to achieve a meaningful distortion, we need to know what *not* to use.

Place yourself before the model, front, back, or side. Run through the gesture on your paper sighting lines, *very lightly.* Then turn your referencing to the edges of the body. Note where lines appear to cross, such as the inside part of an arm resting on the top of a leg. Note where lines converge, such as the shoulder line to the line of the upper arm.

When we speak of looking for angles/intersections, that is what is meant. Look for any of those angles or intersections and draw just the places where those lines meet. Of course you need to remember that the proportions of the body you are crossing are indeed in ratio to something above, below, and beside it. You need to let your eyes scan the body continually so that the proportions will remain convincing. The student illustrations show you several different types of interpretation, all consistent with looking for angles/intersections.

Jack Levine, in his *Boy's Head in Profile,* shows a variation on the use of angles/intersections. He has charted the features and topography of a young man so that ratios, through vertical, horizontal, and diagonal lines, retain accurate spatial distances.

Picasso, at a precocious ten years of age, saw the value of intersecting angles.

***Figure 4.11***  *A drawing showing angles/intersections.*

***Figure 4.12***  *A drawing showing angles/intersections.*

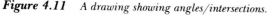

**Figure 4.13** *A drawing showing angles/intersections.*

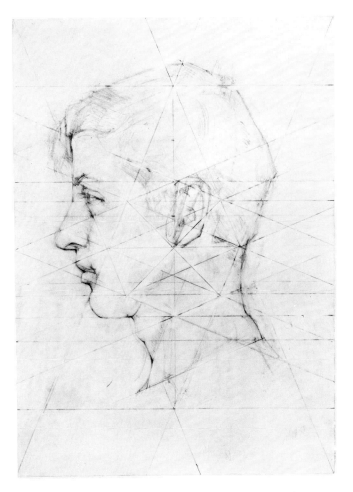

**Figure 4.14** *Jack Levine (1915– ).* Boy's Head in Profile. *Graphite, 358 × 255 m. Courtesy of Fogg Art Museum, Harvard Univ., Cambridge, Mass.. Bequest of Dr. Denman W. Ross.*

**Figure 4.15** *Pablo Picasso (1881–1973).* Study of a Profile. *Conte crayon, 9.3″ × 12.2″. From Barcelona, Museo Picasso. 1892–93.*

# CROSS CONTOUR

## Materials

*2B or 3B pencil*
*18" × 24" white drawing paper*
*Model seated or standing*

Cross contour roams over surfaces. The contour line traverses the topography of the human body, the face, the hair, whatever. Again, press the pencil harder if a shape recedes from you; less so, if the shape protrudes. The line remains a contour line. However, the direction of the pencil can vary. The following example is expressed with a technical grid.

As you can see in the example, there are no edge lines. The horizontal and vertical lines indicate the volume of the figure, but there are no definitive edges finalized with lines. The *edge* of the body is suggested or inferred through the use of multiple cross contours.

The next example shows a strictly horizontal line wrapping the figure. Again, an outer contour is only implied through the use of repeated horizontals.

The following example was done in a somewhat random way but with much care for the way the lines undulate over the form. Within this drawing the student took time to *feel* his way *around* the human figure, which gives this image a sense of mass and depth.

Cross contour can be used for separate parts of the body as well. In the following example, the head was selected. Be sure to note how the student did *not* draw the features as features, but as smaller units which are part of a larger unit. The eyes do not indicate pupils. Remember, you are looking at the figure or its parts as a group of undulating units. Those parts come forward, they lie flat, or they recede. So it is with this example. Note, however, in this example the student used a line on the outer limits of the face to identify an edge.

Ernst Fuchs uses the cross contour in a portion of his *Epic of the Family.* The repetitions of the lines that rise and

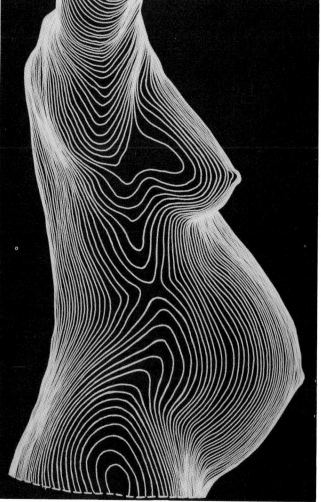

**Figure 4.16** *Andrew Williams.* Lateral Color Map of a Full Term Primipara Produced by Light. *1977. Kodak transfer paper, toned with color center dye. (Collection: Arts' Council of Great Britian.)* The Nude, A New Perspective *by Gill Saunders, p. 134.*

**Figure 4.17** *Horizontal cross contour.*

fall over the body and within the hair help to transform the figure into something ethereal, a symbol rather than a person.

Seated before the model, go through the same sequences you have thus far for the other exercises, drawing *very lightly* for the gesture, the sighting, and the volumes. Just give yourself small or brief indications. Preliminary work at this point is done just long enough so that you give yourself guidelines for the actual problem, which is the cross contour. Overly conscientious students may tend to draw too heavily for the initial layout or use too many lines, which becomes unsightly and extraneous to the problem of cross contour.

When your compositional layout is complete, choose your starting point. Let your eye drift slowly over the physical forms. Coordinate the speed of eye to hand and do not draw faster than the slowness of your eye allows. This work could be fun, if you can strike a balance between concentrating on the physical form and just allowing your pencil to wander with your eye.

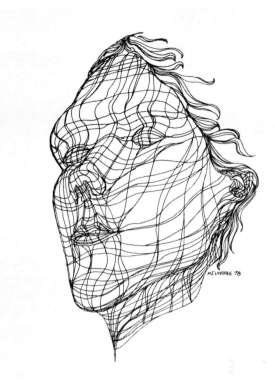

**Figure 4.19**   *Cross contour of a face.*

**Figure 4.18**   *Random cross contour.*

**Figure 4.20**   *Ernst Fuchs.* Epic of the Family. *Mixed media on fiberboard. 1948. Ernst Fuchs. Harry N. Abrams., Publishers, New York. 1977. p. 72.*

# NONSEQUENTIAL FORMS

## Materials

*2B to 4B pencil*
*18″ × 24″ sheet of white drawing paper*

Most beginners are very much aware that their skills are limited, they cannot draw what they have in mind, and that others might laugh or criticize their work. But often their own solutions lead them down a less productive path. The tendency is to want a realistic image. So, the student sets about earnestly trying to draw that way, not fully understanding the components.

Nonsequential forms is a variation on exercises using abstractions of the body. And, as you have probably noted, one of the main ideas behind drawing is learning how to abstract visual clues from what you see. Seeing is as much through your mind's eye as through your physical eye.

As you move through these exercises, trust the process. Centuries of methods have been tested, experienced, used, and taught. Rare is the creature who can jump beyond introductory skills with any success. So, trust the process even though the abstractions may not yet make sense.

With this problem you are asked to draw only parts of the body. The object is to create a composition of parts. By creating a composition of parts of the body, you will infer a figurative element, although the final image will appear strange to the eye. The drawings below include hands, knees, shoulders, elbows, and so on.

One of the best ways to approach drawing nonsequential forms is to sit before the model and draw a section of the body, a portion of a limb, for instance. Get up and move to the back or side of the model. See what other similar shape of the model's body might be placed next to the one you just drew. An elbow/lower arm might be placed in such a way that the calf of the leg becomes an extension of the lower arm. Move to another position and draw again. Continue until your composition is completed. Remember, you can vary scale, direction, and shapes.

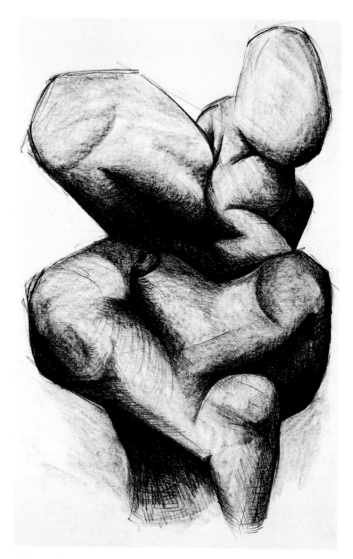

***Figure 4.21*** *Nonsequential forms.*

***Figure 4.22*** *Nonsequential forms.*

# FORESHORTENING

## Materials

*2B or 3B pencil*
*Good drawing paper*
*Draw model from head looking down or from the feet looking up*

Foreshortening is reducing or compressing a figure in a distorted way so that the figure conveys the illusion of receding in space. Foreshortening needs a graphing system with closely focused lines so that the distortions remain consistent. The model should be placed above or below you, or lying down so that you are either at the head or at the feet.

Notice how body shapes overlap in our examples. Overlapping shapes is generally what foreshortening is.

Using the example below, try a small experiment. Place your pencil on the illustration horizontally on the figurative image. Note nearly three-quarters of the image is devoted to the legs. Only one-quarter of the image is given to the torso and head. If you can recognize and trust peculiarities like those and draw them as your graphing lines indicate, the final image will appear to fit together.

As you work through this foreshortening process, be particularly aware of the gesture, the sighting, and the volumes. Draw *through* the body so that you are very much aware of volumes behind other volumes. Try drawing volumes so they appear to create glass units. Preliminary drawing is part of the visual support you need to achieve good work. Try not to let the preliminary work become so profuse with lines that those lines interfere visually with the actual exercise of foreshortening in contour line.

Once you have positioned yourself before the model, lay out compositional margins. Sight height and width. You

***Figure 4.23*** *A foreshortened figure.*

may be surprised how little paper you need to use. In Example 4.22, the model was lying down, so the length of the body was foreshortened. Sight the figure. Is it wider than it is high or higher than wide?

Prior to that, you may want to do a quick blind contour as a small single preliminary sketch, say 6″ × 9,″ to see what shapes actually cover others. Sight as though the graphing lines are close together. Where is the chin relative to the shoulder? Align your pencil horizontally across the model and see. Where does the chin rest relative to the legs or feet? Since you have your model placed, perhaps in a different pose than is illustrated, sight the shapes given to *your* eye. Verticals/horizontals gesturally, volumes next, then contour.

Robert Longo's *Men in Cities: Final Life,* 1981–82, uses foreshortening almost in silhouette. His man is in a stop-action pose. If you measure the length of the foreground hand to foreground foot, surprisingly, they are about the same length. So, do not let your eye fool you. Sight carefully, remembering to sight vertically and horizontally in relatively small graphing units.

The drawing paper has length and height. It does not have depth. The drawing is in two dimensions. You are working with the illusion of three dimensions, foreshortening.

In Jean Ipoustegy's charcoal drawing, the foreground thigh and hip shapes are overemphasized to lend a sexual but somewhat humorous tone to the foreshortened figure. Again, the use of silhouette identifies a contour edge. Looking at these examples you can easily see one shape over another and what the artist did to retain a consistent figurative image. Now it is your turn. Remember, when you finish your work, ask whether the figure you drew could, if it came alive, stand up? Or would the figure be lopsided with one arm or leg much shorter than the other. Would the head appear as a pancake? Giving your image permission to stand up out of the paper before you is a good test of whether you sighted well and understood the volumetric space.

**Figure 4.24** *Robert Longo. Untitled, 1981. 1981–82. MP#D-53. Courtesy, Metro Pictures, N.Y.*

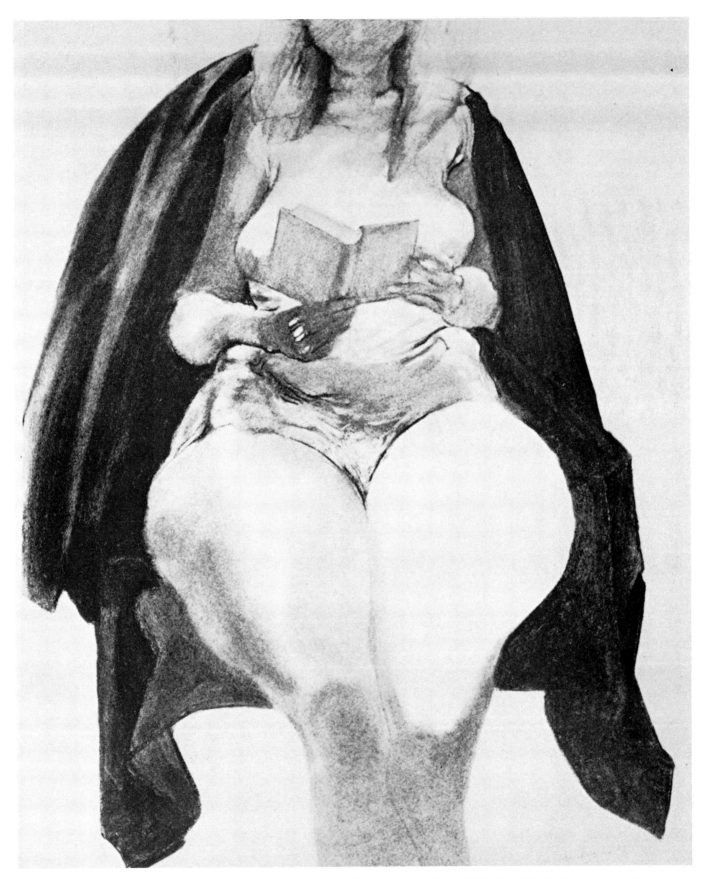

**Figure 4.25**   *Jean Ipousteguy (1920–). Woman Reading in Black Cape. Charcoal, 27 1/4″ × 22″ (69.3 × 55.8 cm). From Galerie Claud Bernard, Paris. 1973, (in* Naked to Nude *by George Eisler.)*

## BREVITY, THE SUCCINCT LINE

### Materials

*Dark pastel stick or brush and ink*
*A quantity of good 18″ × 24″ drawing paper*
*Model in any pose with some drapery*

Brevity in line means more than just hurrying. Information needs to accompany the stroke, and that is not an easy task for speed. To do this exercise really well, practice could continue until each of us grows to a great age. However, most school terms are in semesters or quarters, and an undergraduate degree is usually four years. Practice is the key.

For the *succinct line,* you need to choose the most important line or lines that will convey the most information about a particular body unit. Use as few of those lines as you can with whatever variation of pressure, pastel to paper, might indicate the mass with line only. In effect, you say the most with the least.

Generally, you do much more looking than drawing in this exercise. Observe the model. What is the briefest line you can use to create the image? You might use edges, you might not. You might just sense the movement of the whole figure and draw that. You might sense the flow of the hair and draw six lines.

Look at the drawing of Matisse. Much of his work was of this order, using only the briefest possible lines. He is famous for his poetic line. In *Figure in an Interior,* he suggests the weight of the model resting on her arm—in one stroke, shoulder-to-elbow-to-arm-over-a-pillow. If you study the drawing of Matisse, you will see how masterful he is in drawing a succinct line. The right foot lies on the left foot with one line.

The work in this lesson begins by observing the model. While you are observing, decide on the lines that are the most important to the pose. Then try drawing them with the pastel. If the pastel stick is too stiff for this kind of poetic line, try the ink with brush.

You may use many, many sheets of paper for these exercises. Trial and error processes do take a long time. Expect to try at least 10 to 15 drawings before you begin to understand the complexity of simplicity. But do keep working. Look, assess, draw. Look, assess, draw. When you are done with a drawing, go to another sheet of paper and begin again, either from the same position or a new one. You may get better drawings from another point of view to the model.

Both examples of student drawings suggest and infer the figure through combinations of outer lines, gestural centers of shapes, and body limbs that flow one in, over, and onto one another.

The two examples were done quickly after much observation.

***Figure 4.26*** *Henri Matisse (1869–1954).* Figure in an Interior. *Pen and ink, 26 × 36 cm. From Matisse archives. 1927.*

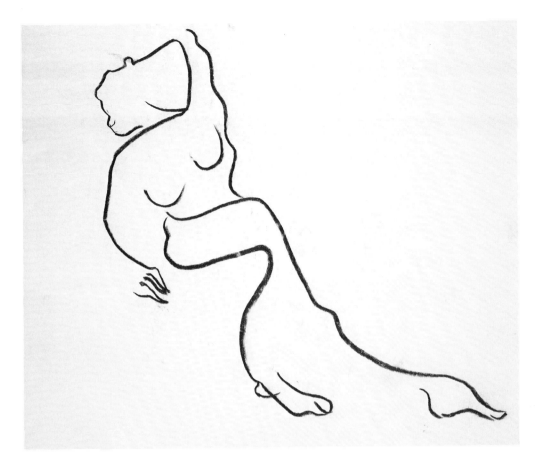

**Figure 4.27**   *An example of fluid line.*

**Figure 4.28**   *An example of fluid line.*

71

# WHEN LINE BECOMES SHAPE

## Materials

*Brush and ink, your fingers and ink, or a very large piece of charcoal (2" diameter) or an Oil Paint stick or cattle marker (1 1/2" dia. grease base stick)*

*5–10 sheets of good drawing paper*

*Model assumes any pose*

When line becomes shape usually means that the line is wide enough so that strokes you make on your paper appear as a full form such as a hand. Note Richard Stankiewicz's *Untitled.*

His use of ink on paper appears to be a hand. Its spontaneity conveys the mass, motion, and tension of a hand. The brevity of the line could refer back to the succinct line, but here the line is big and wide enough to become an entire shape of its own.

In Stankiewicz's work the hand is dark. The opposite of that, light shapes, can work too. Strokes surrounding a human form can make the figure appear as though done in one or two lines. Negative space becomes positive. The *nonstroke* is what is left and creates the figure. Note Dorothea Tanning's *High Tide.* Her watercolor and gouache on paper does exactly that.

The student illustrations below suggest ways of working with line-into-shape. The following example indicates a

**Figure 4.29** *Richard Stankiewicz.* Untitled. *Ink on paper, 17 1/4" × 8 1/2". From Zabriskie Gallery. 1960.*

**Figure 4.30** *Dorothea Tanning.* High Tide, 1972. *Watercolor and gouache on paper, 8 3/4" × 10 1/2". Kent.* ArtNews, Summer 1987. *p. 212.*

seated woman. There are some fine lines indicating face, upper right arm, and left hand and toes. But in one stroke, the student indicates hair. In another, the right side, bottom and thigh. In two loops, the breasts. Two more lines indicate the model's right arm and leg. Six large lines indicate the mass of the figure.

Another student chose to work with her fingers in ink. She selected the model's head to use for her drawing. The lines here, most especially the horizontal and diagonal within the facial image, indicate the brow and nose respectively. The strokes are minimal, dramatic, and large.

To set about working on this exercise, decide whether you want to work with the full figure, or part of it. Select your media, preferably wet. Positive and negative space will be prominent in your drawing. You have several illustrations that suggest possibilities for drawings of line-into-shape. Use dark media to suggest the body or body part you choose. Or, use a negative white to infer a body shape. Using external lines or gestural strokes wide enough to make full shapes are choices you can incorporate. Remember, these are suggestions. You might find combinations of the above approaches or new ones of your own.

***Figure 4.31***   *Student drawing, line into shape in ink.*

***Figure 4.32***   *Student drawing, line into shape.*

# FIGURE INTO LANDSCAPE

## Materials

*Charcoal or pastels*
*White drawing paper*

Figure into landscape is not a new idea. Many artists have taken the human figure and constructed it into something that it is not: a landscape, vegetable, or whatever becomes anthropomorphic, carrying human characteristics. Andre Masson, in his *The Cascade* drew such a picture of lovers into mountains. His drawing is done in contour, without much value or shading added.

The student illustration is a gesture. If you look closely you can see the original drawing is from the figure, a woman, lying in such a way that her back is on the table, her hips turned toward the viewer. By making the curvilinear lines of her shape into topographical lines of land forms, adding some vegetation and rock formation (which was a blanket on the table) behind her, the image is transformed into a landscape with figurative elements left—vestigial figurative elements.

You might want to try this problem with the figure lying down among some pillows or rumpled blankets. I would suggest that you use a short piece of charcoal to get the linear masses of the human figure. Thereafter, turn the charcoal on its side and use the width to shape a landscape from the forms you have drawn. Because the linear drawings can be done quickly, you might want to have the model pose four to five times for five-minute drawings; you can then work on the figure-in-landscape transformations while the model takes a break. ▲

**Figure 4.33** *Andre Masson.* The Cascade. *Ink, 46 × 35.2 cm. 1938.* Naked to Nude, *by George Eisler.*

**Figure 4.34** *Student drawing, figure into landscape.*

# 5
▼

# Value/Mass/Texture

## LIGHT SOURCES—VALUE

Following a brief discussion of light sources, we will suggest exercises through which you may practice toward achieving the illusion of three dimensions, or mass, in your drawings.

Without light we could not see. The simulation of values, which are gradations of light against objects, has been with us for thousands of years. We can see forms recede because light and shadow play across surfaces. Lights and darks typically finish a drawing, make it seem complete. Several kinds of lighting should be considered. One is *Natural light,* for the sun or moon. Another is *Artificial* as in light bulbs or candles. *Holy light* (for lack of a better term) is assumed to come from something holy or ethereal. *Luminous lighting* is a made-up atmospheric system of lights and darks ordered throughout a drawing or painting.

You have probably seen pictures of a sphere, perhaps a ball, indicating where the highest light is, the next shade, and so on through each value until the darkest of the shadows is identified. The ball was lit, using natural light from a window, perhaps. That is one kind of light source used by artists.

Simon Faibisovich's *Local Mad Woman,* 1987, is a very good illustration of the use of *natural light.* The woman enjoys her "piece of the sun" on a very crowded sidewalk with preoccupied passersby. As you look at the image, you

***Figure 5.1*** *Simon Faibisovich.* Local Mad Woman. *1987. Oil on canvas, 50 1/2″ × 56″. Phyllis Kind Gallery.*

note that the sunlight comes from the upper right area. All of the light shapes are to the right in the painting, leaving areas of darker tones scattered variously throughout, but weighted to the left in the painting.

The student illustration shows natural light coming from two sides, the right and the left. The model was posed before some stacked items on a window ledge. The light from the left highlights his hair, a portion of his face, his right shoulder, and his chest. This light outlines the top of his right arm, his left knee, and lower leg. The light from the right highlights the left side of the head of hair, which falls forward over the face. This light is apparent on the model's back, his rump, his left thigh, and leg.

*Artificial*, man-made lighting comes from light bulbs, fires, spotlights, fireworks, and so forth. The following three examples indicate some of the variations one can attempt when working with man-made lighting in composition. The three illustrations use lighting as a vehicle to enhance singular dramas. *The Fool and the Maiden*, by Werner Tubke, shows a woman in a white silhouette (that in itself is strange since most silhouettes are understood to be dark) against a very dark unidentified background. The artist uses light (in this obscure allegory, an elaborately dressed fool pointing to a woman removing her clothes) which spotlights both figures in the foreground, the fool in general, the woman in particular.

Lighting in our next illustration, Randy Hayes' *The Price You Pay*, seems to be a searchlight from the right, shining on two figures standing in the dark. There is little

**Figure 5.2**  *A figure drawing showing two light sources.*

**Figure 5.3**  *Werner Tübke*. The Fool and the Maiden. *Oil on canvas, 59″ × 59″. 1982. Courtesy Gallery Schlesinger, NYC.*

natural atmosphere in his work, but lighting suggests, along with the title, a heightened psychological atmosphere of fugitives being apprehended, or a clandestine rendezvous discovered.

Another form of man-made lighting is suggested by David Siqueiros in his *Echo of a Scream*. Because the image identifies a child painted silver grey, sitting in the midst of demolished surroundings painted in greys, blacks, and reddish browns, one can assume the light comes from detonations of dropping bombs. The light source seems to come from above and left, casting shadows to the right. The drama of the child is heightened because a second image of the baby's head looms above and within the smaller body, magnifying the screams of all children of war.

A more contemporary illustration using *Holy Light* is that of Ernst Fuchs in his *Mary with the Infant Jesus*. The Christ Child is bathed with light from within while God the Father radiates shards of light behind the Madonna.

**Figure 5.5** *David Alfaro Siqueiros (1896– ).* Echos of a Scream. *Enamel on wood, 48″ × 36″. Collection, The Museum of Modern Art, New York. Gift of Edward M. M. Warburg.*

**Figure 5.4** *Randy Hayes.* The Price You Pay. *Pastel on paper, 53 1/2″ × 34″. 1984.* ArtNews, *May 1986. Randy Hayes/ Collection, John H. Noonan.*

**Figure 5.6** *Ernst Fuchs.* Mary with the Infant Jesus. *Resin oil glaze and gold leaf on wood, 39 3/8″ × 39 3/8″ (100 × 100 cm). 1960–62. Ernst Fuchs.* Harry N. Abrams, Inc., Publ., *New York. p. 125.*

*Luminous light* is most often associated with a diffused phosphorescence throughout a drawing or painting. Cubist theory included, among other concerns, atypical lighting. In Pablo Picasso's *Ambroise Vollard* one cannot see a consistent source of light. Light is there, but really has no specific source.

The student example shown below indicates naturally cast shadows beneath the breasts, over the model's right thigh, and below the left leg. But for the figure and draperies, lighting was made up.

Selina Trieff uses a diffuse and phosphorescent lighting in her drawing called *Duo*. There is no real source of light here either. The picture is simply imbued with light shapes to emphasize the meaning and content of her very dynamic work.

**Figure 5.7** *Pablo Picasso (1881–1973).* Ambroise Vollard. *36″ × 25 1/8″. Pushkin Museum, Moscow. 1909–10.*

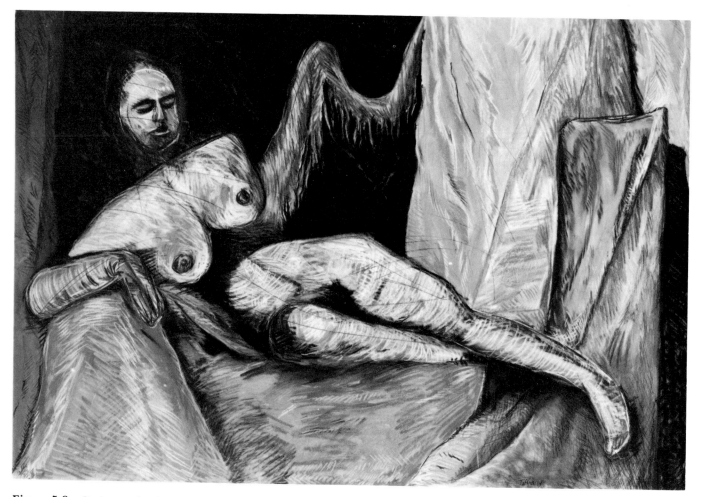

**Figure 5.8** *Student work using combinations of lighting effects.*

***Figure 5.9*** *Selina Trief.* Duo. *1984. Charcoal on paper, 72″ × 60″. Courtesy Graham Modern Gallery.* Arts Magazine, *Summer 1987. p. 100.*

# CROSSHATCH

## Materials

*2B pencil*

*3B and 4B pencils, if necessary*

*White or off-white paper*

Crosshatching has many variations. Simply put, one line crosses another. If you stroke the paper with a series of two- to three-inch diagonal lines from upper right to bottom left, then stroke over those same lines from upper left to bottom right, you have created a crosshatching line. The lines can be very orderly, using a ruler or compass. The lines can be straight or arched using no other guide but your eye. The lines can be scrambled to some degree and still be considered crosshatching. The lines can be thin or thick, dense or sparse, and can be combined with other kinds of shapes and lines.

The intention behind crosshatch is to build volumes without using one specific line to identify edges of figures. You will find that erasing a crosshatch is more difficult than a single line. The crosshatch is a nest of lines. Erasing back is almost an impossibility without major repair or re-drawing of a large area. Edges are indicated through the repetitions of lines. Look at the example by Hendrik Goltzius' in the detail of his engraving, *Hercules Victor*. The lines are very consistent and follow the topographical forms of the body. If you look closely, the edges of the figure are visible because enough care was taken to place one line after another in such a way that the beginning point of each line has created a simulation of an edge. As Goltzius worked, he made thicker lines more dense and thinner lines less dense to indicate values.

Pablo Picasso in *Tete de Femme*, a drypoint, a type of printing different from Goltzius' engraving, uses dense and rounded lines for the hair atop the head with less dense and straighter lines for edges along the cheek, eye, mouth, nose, and throat. The hair has an edge developed exclusively by the crosshatching method.

Both student illustrations incorporate a more reserved crosshatching than that seen in the Rainer work. The image in the following example is difficult to discern for some viewers. A model is seated on a table covered with material, her leg extending left. Windows are behind her. The student did not *push* or refine the drawing so that edges could be seen clearly.

The second student work is drawn in such a way that the image is more defined, although the edges of the figure are indicated through crosshatch and not contour.

The best way to begin a crosshatch is simply to make crosshatching lines over the whole paper. Turn your paper as you continue making the relatively short lines overlapping each other. This procedure allows you to build a surface with which you ease into the actual crosshatch drawing.

**Figure 5.11** *Pablo Picasso (1881–1973).* Tete de Femme. *Drypoint, 27 1/2" × 20 1/8". Achim Moeller Ltd., 8 Grosvenor St., Bond St., London, W1X9FB. F.A. Library, Leicester Sq. #6.*

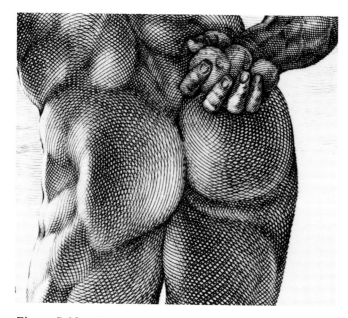

**Figure 5.10** *Hendrik Goltzius (1558–1617).* Hercules Victor. *Engraving, detail. 1617.*

**Figure 5.12** *A crosshatch drawing using natural lighting.*

Fill the page with crosshatch lines as indicated in figure 5.14 on page 83. When you have finished, look at the model and his/her surroundings. Squint your eyes. Squinting helps identify light and dark shapes more easily. The next step often seems difficult for some students, though it should be a continuation of your earlier strokings.

Scan the whole area you have chosen for a composition. Keeping your pencil working in crosshatching, look for the lightest *shapes*. Disregard objects as objects. Look for the lightest *shapes*. Crosshatch those light shapes covering the whole paper with shapes of light that are a similar value. Then look for the next darkest series of value *shapes*, middle-light shapes. Keep your pencil moving, crosshatching with a little more pressure for the middle-light tones. Continue to the next darkest series of value shapes, into the middle tones, and on down the value scale.

Remember to scan the *whole* set-up, looking and drawing *like-minded value shapes*. If you linger over objects, if you just draw singular pieces of things, you begin to identify edges, and a sensitive crosshatch drawing can easily become dull and uninsightful.

***Figure 5.13*** *A crosshatch drawing using a man-made light source.*

**Figure 5.14**   *Crosshatching lines over the whole page.*

# ERASURE DRAWINGS

## Materials

*Soft vine charcoal*
*Kneaded and plastic eraser*
*18" × 24" newsprint pad*

Erasure drawing means using the eraser as the drawing tool. To do that the paper needs to have soft vine charcoal already stroked or smeared across the surface. Again, the surface is built first before you begin to draw, just as in crosshatching.

Because this problem can be as short as thirty minutes or as long as three hours, the model should be placed in a comfortable position. The room should be darkened and a spotlight, or dramatic daylight should be used on the model and the environment.

With crosshatch you went from light shapes down to darks. Here we reverse the tone of the paper, pulling from dark back into light. Observing the student illustrations, note how the student *erased back* to the lightest shapes first. In other words, erasing charcoal off the paper will indicate light shapes. Squinting your eyes helps to isolate the highest highlights and the darkest darks.

As you learn to draw, exercises will lead you to expressive work. Try not to distort the exercises into something more comfortable that allows you to fall into some older, easy habits. You could miss very important learning experiences.

Let's start you on your way. In your first drawing the very best thing you can do is cover the paper with soft vine charcoal, twist your kneaded eraser into a shape about the circumference of your little finger, squint your eyes, *seek out the lightest shapes throughout the whole composition and draw those shapes.* Forego drawing individual things, such as the model, the table, the drapery. By seeking out the lightest shapes throughout the whole environment, which includes the model, and erasing those light shapes from the vine charcoal field on your paper, you begin to see compositional space as a unit. If you draw that way, erasing the lightest, next lightest, middle, and middle dark tones *throughout the whole composition,* you teach yourself to see with an artistic eye. If you continue to draw individual objects, quite probably your skills will develop more slowly and eventually will be less fulfilling. Andre Agassi is known internationally for his gifted tennis playing. He says that his intention is to play the best that he can play, and "the wins follow."

***Figure 5.15*** *An erasure drawing.*

Drawing is like that. If you work for the structure of drawing, the pictures follow. Part of learning structure in drawing is isolating one aesthetic issue to practice at a time. This problem works with lights and darks simply as shapes in themselves. If you select only light and dark shapes, not objects, the result will be a picture of a model on a model stand, and you have taught yourself to take one more step into independent work. Drawing just the light shapes with the eraser, then adding more charcoal for the darkest shapes, is the best possible way you can work within this exercise.

**Figure 5.16**  *Bill Wilman.* Heat from a Black Sun. *1984. Oil on canvas, 78″ × 66″.*

## THE MIRROR IMAGE—DISTORTIONS

### Materials

*Charcoal, ink, or pencil*
*18″ × 24″ good drawing paper*

Many, many artists have used distortion as a device for content or meaning in a work of art. Often the distortion is a simple reflection in a bent mirror, which is what our exercise will be. Sometimes the distortion is a variant on linear perspective. Purposeful compression or elongation are also approaches to distortion. Huge figures in a small space, another. Magritte was famous for making pipes, combs, and shaving brushes larger than the bedrooms in which those items rested. Relationships change. Natural proportions appear false. The distorted image invites a response stimulated by proportions that appear nightmarish, repugnant, or impossible.

Bill Wilman elongates and compresses his very large portrait, which is over six feet both in height and width,

yet his distortions of the head and face do not deny nature. The people whom he paints are still recognizable.

Fernando Botero tends to work on very large canvases or paper, sometimes 12′ × 13′. His *Mona Lisa, Aged Twelve,* an early work, is a monumental doll-like figure of a girl. Sometimes his figures are wider than tall.

Saul Steinberg chose another type of distortion in his *Canal Street Battle.* The figures have disproportionately long legs. This form of distortion moves into caricature, which often uses satire or mockery through comic types.

Anthony Green gives us an overhead distorted, three-point perspective of a room with three figures inside. The title, *My Mum's Dream,* suggests either her hope or her actual dream, lending credence to the disjointed space.

The student examples indicate the use of bowed mirrors, an image that is bent for you. The mirrors used are two large pieces of mylar-backed plastic, one concave, one convex. Placed on the table with the model seated very close to the point where the mirrors come together and form an angle, distorted reflections can be seen from a large area

**Figure 5.17** *Fernando Botero (1932– ).* Mona Lisa, Age Twelve. *Oil and tempera on canvas, 6'11 1/8'' × 6'5''. Collection, The Museum of Modern Art, New York. Inter-American Fund.*

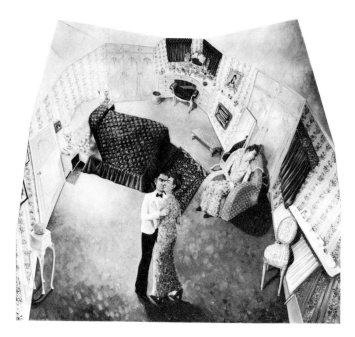

**Figure 5.19** *Anthony Green.* My Mum's Dream. *1974. Oil on board, 88'' × 95'' (223.5 × 236 cm). Mayor Rowan Gallery Ltd., Anthony Green. Arts Council of Great Britain. #1.*

**Figure 5.18** *Saul Steinberg.* Canal Street Battle. *1987. Oil pastel on paper, 59'' × 83 1/2''. Steinberg. Pace Gallery, 32 E 57 Street, New York, 10022.*

**Figure 5.20**  *Drawing of a distorted image.*

in the room. Obviously, you must stand somewhere within the viewing range of the mirrors.

The model is seated close to the mirrors in order to get as much of his/her body reflected as possible. Because the mirrors are bent, repetitions of parts of the model's body are seen in nonsequential ways. Interestingly, the curved linear striations are Venetian blinds.

The second illustration was taken from a highly polished metal cylinder, a fire extinguisher, actually. The base of the cylinder had another reflective metal piece pressed into it, thus the second image below is not unlike a predella, meaning the smaller picture panel beneath a large altar painting.

Of course, the assumption always is that you first gesture your layout, laying in sighting lines, then drawing simple volumes, before beginning the individual exercise in earnest.

**Figure 5.21**  *Drawing of a distorted image.*

## CREATING THE POSITIVE VIA THE NEGATIVE

### Materials

*Use pencil or charcoal with appropriate paper*

Working with negatives means working with space surrounding or permeating objects, in our case, the figure. As you view a figure, negative shapes could be the areas between the arms and the body, that is, the holes. Another negative would be the area around the outside of the body. If you are viewing a model from the side, sitting forward on a chair, the negative would be the space created between the legs of the model and the chair itself, and so on. The following illustration indicates the negative shapes of the model seated on the floor on a blanket. The most obvious negatives are indicated with her right arm in the shape of space between her forearm and face and between her underarm and thigh.

However, you will also note that in the space surrounding the model, there are items, like a chair. The student drew negative shapes of the chair, bringing some lines of the chair just to a point where the model's hair or her back would begin. By stopping at these points, the image of the model is evident. But a strange thing happens. Since most of the information drawn here is so very flat and so very white, is the figure more, or less, negative itself? Generally, the *positive* form is understood to have the appearance of volume. In our example, the volume of the figure is limited to the use of line by the student, and has no value. But if you study the picture for any length of time, the figure begins to slip from negative to positive to negative.

John Graham's *Woman Seated on a Marble Block* is a case in point. The figure is understood through its light shapes, made possible by the dark shapes. If it were not for the darks, one could not see the lights. And because the lights are placed the way they are, you can see and understand that there is a seated woman. More active shapes are given to the surface of the marble rock on which the woman is seated than to the surface of the woman's body. To create her image the positive figure mass is thrust from negative shapes, made dark, made from shadows.

***Figure 5.23*** *John D. Graham.* Woman Seated on a Marble Block. *1927. Oil on canvas, 39 5/8'' × 25 5/8''. National Museum of American Art, Art Resource, N.Y.*

***Figure 5.22*** *Positive shapes via negative shapes.*

Our third example uses the light and dark shapes that surround the figure. The lights and darks identify the set-up, turning the figure into a very white, flat shape with the exception of the buttocks, where, with a little bit of shading, the rounded forms take on some appearance of life.

Another student used combinations of positives and negatives to formulate new negatives and positives. The visible parts of the figure were shaped first by the background—darkening the Venetian blinds behind the head, wall, and shadow drawn from the neck down. The residual light shape of the figure was modified then with values so that the model's right arm remains flat while the model's left arm, chest, neck, and face carry just enough *value* to infer mass.

You can play alternately with this problem as you see new options. What is sometimes a negative becomes a positive, and vice-versa.

**Figure 5.24**  *Drawing the negative to achieve the positive.*

**Figure 5.25**  *A positive-negative drawing.*

# TONAL RANGES/DIFFERENT PAPERS

## Materials

*Light-toned paper & 3 light-toned pastels OR*
*Middle-toned paper & 3 middle-toned pastels OR*
*Dark-toned paper & 3 dark-toned pastels*

Another important issue in drawing lines is perceiving value ranges and comprehending what they can do for you. This exercise takes you through a series of tones in paper and media. Our scale of lights to darks ranges from white to black. Within that scale a person can select light paper and light-toned chalks, middle-toned paper and middle-toned chalks or dark paper and dark-toned chalks. The harmonies created in these works are delicate. With four ranges of tones such as white paper, off-white, deeper white, and creme pastels, or middle gray paper, middle gray chalk, darker gray, and dark middle-gray chalk, you begin to see why the works are subtle.

Leon Golub has several works that incorporate these specific ranges of values, using his themes of mercenaries, riots, and victims. The lightest scale can be seen in *Horsing Around III*. The woman, background, and clothing almost fade one into the other, they are so very light. Only the skin tones here appear somewhat darker than the rest of the tonal ranges. The mercenaries are at play.

Golub's middle-toned work, *Mercenaries I,* shows middle-toned ranges throughout most of the picture, the face of the center figure being the exception. The mercenaries wait.

The darkest values in his *Interrogation II* sustain and underline the act of interrogation. Figures are close to and parallel to the picture plane. Clothes and parts of the figures are lighter than the background. Golub's backgrounds are usually stark and flat, often bright red like the grounds one sees in the Pompeiian wall paintings. The rank simplicity of the color sets an extraordinary visual stage wall for the activity just before it.

*Figure 5.26*   *Leon Golub.* Horsing Around III. *1983. Josh Baer Gallery, NYC.*

*Figure 5.27*   *Leon Golub.* Mercenaries I. *1979. Josh Baer Gallery, NYC.*

*Figure 5.28*   *Leon Golub.* Interrogation II. *1981. Josh Baer Gallery, NYC.*

For each of the following examples choose the paper tone—light, middle, or dark—with which you want to work. That choice sets your value scale. Then select like-toned pastels, meaning chalks. If you opted for light papers, then select three very light, close values of the same hue for the pastel sticks. If you prefer the middle- or dark-toned paper, choose the pastels accordingly.

Once the model is posed, begin your drawing with the lightest colored chalk to make the gesture, the sighting, and the volumes. With that much structure laid out on the sheet for composition, the following steps should be relatively easy.

The values one normally sees within a set-up range from lights to darks. Because you have a toned paper and three close values, the whole composition needs to be transposed to four values, those of the chalks and that of the paper. For instance, if you select light paper/light pastels, the darkest dark of the model/set-up will be the value of your *most dark* light pastel. Codify or transpose all of the values into the four tones. Remember, of course, that the paper can and should be used as one tone. You might find an interesting variation for positives and negatives in this problem.

***Figure 5.29*** *A drawing using light paper and light pastels.*

***Figure 5.30*** *A drawing using middle-toned paper and middle-toned pastels.*

***Figure 5.31*** *A drawing using dark-toned paper and dark-toned pastels.*

# FINGER DRAWING

## Materials

*Oil-based modeling clay*
*Various drawing media; that is, charcoal, ink, pastels*
*White drawing paper*

Finger drawing can be done with many media—ink, chalks, and clay are just a few. The intensity of the finger-stroke carries a power in it that is unusual because you are literally stroking the paper with your flesh. The tactile sensation of finger with media onto paper is more immediate than with any other drawing utensil. It is pleasureable. However, much of the pleasure in the finger-drawing exercises comes at the expense of control. With media on your fingers, your stroke area is the width of your finger, and making thin or thick lines almost becomes a moot point. Still, you can use your fingernails for thinner lines.

Let us start with a picture using modeling clay. A backing board of a middle or dark tone would be helpful if your clay is light, or a light board if your clay is dark. Have the model take any pose, for about ten to fifteen minutes. Stand over the board at a table if you can, or before it on an easel. Warm up some of the clay by rolling it in your palms so that the clay will smear easily. Look for the structural lines in the figure. Begin by putting a small blob of clay on your finger and start drawing. You will be surprised how easy working this way is, if the clay is soft. If the clay is cold and hard, your efforts will be frustrating. With this kind of drawing, and with the brevity of time, making a bas relief (a sculpture that rises part way from the flat background) is not the intention. Drawing with your fingers suggests and relies on spontaneity. You can see in the illustration the student has drawn with the light clay and has some charcoal mixed in to indicate darker shapes.

Keeping a small ball of clay in the palm of your non-drawing hand keeps the modeling clay warm, soft, and accessible, since each stroke probably means you will need a chunk to put on your fingers in order to draw the next stroke.

***Figure 5.32*** *A student applying charcoal to an image drawn with clay.*

93

**Figure 5.33**  *A gestural drawing using clay and charcoal.*

**Figure 5.34**  *A gestural drawing using clay and charcoal.*

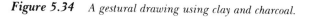

Drawing with charcoal as a medium poses some special problems. You need to pulverize enough compressed charcoal, several sticks, to put into a small container, like a small cup. The charcoal does not go as far as the clay. Dipping a finger into the pulverized ash puts the medium on the tip, but the strokes tend to bleed off very quickly since charcoal does not spread as easily as clay. A 9″ × 12″ piece of paper would do fine for this exercise using charcoal as the medium. Much larger than that and your strokes get too large and too long. Long and wide strokes will literally burn your fingers. And the strokes certainly use up much more charcoal.

Combinations of drawing utensils and fingers are appropriate. In the next example, a student drew a gestural study of a seated female figure seen from the back. Some of the strokes are made with a pencil swooping through pools of India ink previously applied with fingers.

Certainly, drawing with fingers can be taken to more complete images than the simple gesture. Chuck Close's work, *Georgia, State III*, is an etching done with fingerprints, not strokes with fingers, just fingerprints. But it is

an example of what artists can do when they set their minds to something quite unusual and out of the ordinary.

Drawing with ink and stick is included under finger drawing because as you draw with the stick in your hand, you can include working with your fingers as well. Find a stick on the ground somewhere. You might want to try two or three sticks and see which ones you like best. A bottle of India ink, or sepia ink, or both, can be used. A sheet of good paper, perhaps a printing paper, a sheet of watercolor paper, or a more absorbant and less smooth drawing paper can be used.

The student whose work is illustrated below used several small twigs he found under some of the campus trees. He had *washed* his paper with pale, thin colors of blue and green ink before he came to class. He had prepared several papers that way. Washing in this case means using a moist sponge, dipping the sponge into a watered-down colored ink, then wiping that ink across the paper with the sponge. The shapes he made were spontaneous shapes he let dry on the papers in order to bring them to class. Once in class he could work on the prepared papers with a stick, ink, and fingers to draw his images.

**Figure 5.35** *A mixed-media drawing using fingers, ink, and pencil.*

**Figure 5.36** *Chuck Close. Georgia, State III. 1985. Gravure etching, 30″ × 22″.*

**Figure 5.37** *Paper prepared with ink wash, over which an image was drawn with stick, ink, and fingers.*

Robert Arneson's portrait of Jackson Pollock in acrylic and oil stick on paper includes handprints and the stroking of the surface with frenetic energy. Donald Kusprit's article, *Arneson's Outrage,* suggests the methodology used supports the subject of Pollock as a "brutalized victim-survivor . . . 'suicided by society.' " Because of the energy of Arneson's strokes, this portrait is immediate and powerful.

**Figure 5.38** *Robert Arneson.* Portrait of Jackson Pollock #3 *1983. Acrylic, oil stick on paper. 41½ × 30 in. Courtesy Trumkin Gallery. Photo by Eera-inkeri.*

# MIXED MEDIA

## Materials

*Sundry media—ink, pastels, graphite, markers, watercolor, brown paper bags, glue, transfer materials, etc.*
*A heavy sheet of drawing paper, or watercolor paper, or printing paper*

Mixed media means the mixing of different types of media and/or materials. Again, the work remains a drawing, not a bas relief. Our use of mixed media will remain relatively flat. Jasper Johns' charcoal and pastel is an illustration of mixed media with minimal use of the figure. The work was done by dragging a charcoal powdered rag over paper laid on rough ground. Notice the title, *Diver*. That gives you a clue to the image. The diver's hands are at the bottom of the drawing. A motion arcs out and up. One sees the palms again, about the same time that the bottom of the feet come into view. You can sense a figure has dived into deep water. The sensory feeling brings back associative values we have of diving into pools, hands over head. Having dived in, the force of the water past our bodies allows us to pull our arms out, up, and back to our hips for thrust and glide.

Jim Mesple's visionary *Paulette as Venus X-Raying the Trojan Horse* combines a woman with rubber arms to a contemporary Chicago skyline. He cavorts between myth, architecture, and history, past and present. The viewer is incited to finish the implied narrative. In this work he has used charcoal, ink and gouache on paper.

Using resist, embossing, and collaging in his *Chairman of the Board*, Bruce Thayer began building his drawing. He attached additional pieces of paper as his work progressed. The main figure is cut out of cardboard, collaged, glued onto the paper, then rolled through a printing press. The large black areas are shapes of black paper cut out and applied. There are copies of engravings from old industrial manuals incorporated. Thayer used rubber stamps, crayons, colored pencils, and watercolors in his drawing. *Chairman* is one of a series called *Disasters of Industry*. There are three disasters for Thayer, domestic, environmental, and economic. Each of these issues describes part of the human condition.

**Figure 5.39** *Jasper Johns. Diver. 1963. Oil on canvas with objects, 1962. Mr. and Mrs. Victor W. Granz, N.Y. The Leo Castelli Gallery.*

The student illustration shows the use of transfer, xerox, brown paper bags, watercolor, and graphite. The model was placed on a stand at about eye level. Her head and neck rested below the corner of the stand allowing her hair to fall straight down. Her feet were lifted up to rest on a backdrop above the stand. Shapes of other items in the classroom behind her were used as *value* shapes. The other figurative image is another student's partial outline. Our student artist nestled some very famous artwork into the hair of the model.

Your assignment now is to explore the capabilities of mixing different media. Stay within the media you have used to this point, exploring with brown paper bags and

**Figure 5.40** *James Mesple.* Paulette as Venus X-Raying the Trojan Horse. *1986. Charcoal, ink, gouache on paper, 22 1/2″ × 18″. In the collection of P. and E. Solow, Chicago.*

transfers, perhaps. One student brought a drawing on which he had used shaving cream and soot. While the drawing smelled good, the novelty soon fizzled, as did the shaving cream.

Always, of course, one needs to remember composition, gesture, angles, and volume. Once that part is done, review the media and materials you brought, deciding what you want to use for what reasons. For this stage of your work, the media and materials probably should follow formal concerns. For instance, squint your eyes, looking for light and dark shapes throughout the composition. Work your way through the whole composition looking only for dark shapes. Tear those shapes out of a brown paper bag and glue them onto your drawing paper. Limiting the dark shapes only to obvious objects such as the dark hair of the model, is not what you need to seek. Look for dark shapes as they appear throughout the visual area, the whole composition. If a cast shadow from the neck is the same value as the hair, those two things become one shape, or one piece of brown paper bag. So, the questions to ask yourself are, where are the lines, the verticals, the horizontals, diagonals, and curvilinears? What media can I use to suggest those lines? What geometric shapes are there and what media can I use to indicate those geometric shapes in the drawing? What media can I use to push the space back and bring forms forward? And so on.▲

**Figure 5.41** *Bruch Thayer.* Chairman of the Board. *1988. 60″ × 72″. Mixed media on paper. Artist's Collection, 1515 Kelly Rd., Mason, MI or Zaks Gallery, 620 N. Michigan Ave, Chicago, IL 60611.*

*Figure 5.42*    *A mixed media drawing.*

# 6

▼

# Adopt An Artist

In *Some Spatial Options*, Chapter Three of this section, you reviewed broad categories of artistic space, standard spatial *isms* related to the Renaissance picture plane. Those *isms* were Idealism, Realism, and Surrealism. Other uses of space included overlap, equivocal, contradictory, simultaneous, and atmospheric perspective.

This chapter emphasizes more specialized theories by masters of space. Each master artist defined his beliefs around a largely unarticulated experience. Artists tend to paint a speechless reality. More often than not, the artist does not write about his or her beliefs. Usually viewers deduce theories in the art works. Viewers include art historians and critics who write about the subjective space embraced by artists.

The object of these studies is twofold: *first*, to show you viable spatial alternatives to the skill studies you are learning (which are principled in the Renaissance window), *second*, to allow you to adopt a unified theory or method for a short while.

Skill can take you to the heights of copying, sometimes aligned with realism. Skill alone cannot move you to your own stroke, to your unique thumbprint of expression. Adopting other methods challenges a developing artist to work with space in ways that otherwise might not be examined. Most people come persuaded that realism means copying, which, to them, is art. They want the skills to recapture that space. The result rather is a derivative art, one that can only reproduce rather than express. Figures drawn within this outlook may have a representational form but remain artificial and frequently have as much expressiveness as a medical illustration. Growing into your expressive capabilities takes effort alongside a willingness to change, to develop. Some processes work, others do not.

Before we begin our discussion about particular artists, let me offer a suggestion. Work to acquire enough information about a movement or an approach to apply the theory as you are drawing. You will learn very little if you just copy an artist's picture from a book.

Using the fine art section of a library or reading room, try to select well-known, established fine artists that use the figure. The artist or movement should be represented in the book with enough color reproductions so that you can explore the drawings and/or paintings in such a way that the theory is meaningful. For instance, if you want to study Seurat's pointillist method which includes using dots of pure color next to each other, studying his charcoal figurative works will tell you nothing.

Remember, too, that artists generally keep evolving. Most artists change their styles as they grow through life experiences such as wars, marriages, deaths, births, and so on. Select a period of time that demonstrates consistency in the works. De Kooning's lifetime work would be impossible to subsume under one set of operative principles.

Once you find the paintings or drawings of an artist, whether they work singularly or with a group, read about the theories behind the work, writing down the most important points. Those points relate to *the ways the artist used space*. For instance, if you concentrate on adjectives that describe a viewer's response, words like "lovely," and "dynamic" and "powerful," there is little to understand.

Biographical data is not helpful either. Cut to the chase. What is the theory and what artistic devices made it work? How did he/she use line, value, color, texture, shape, space, and composition? The artist as human being with a personal history is for another time.

We will begin our study with a Frenchman, Georges Seurat, 1859–1891.

Seurat was interested in scientific principles of pure color which, when closely juxtaposed, would merge into another color through the seeing eye. The methods he used were these: preparatory studies, a monumental scale, and systematic dots of color used in varying degrees, densities, and colors to interpret a naturalistic setting. The end results were paintings of elegant and delicate gradations of luminous color and light.

For practical applications of Seurat's theory, we need to look at his work in color reproductions, study sections of them, test out materials and paper. Let's describe what Seurat did with each of the formal elements.

**Line:** Seurat executed many preliminary drawings before he started painting, working out his compositions. Line was used to guide his painting process. In his finished work, lines can be seen separating objects, but because of the application of the small dots of paint, the lines are not harsh.

**Value:** Value is very important to Seurat's work. Although he used only blue or color combinations with blue in shading and shadow areas, no blacks, he stayed with natural appearances of light, using sources that cast shadows. The light in our illustration moves from the right to the left of the figure.

**Color:** Color is of primary importance. Using single points of red, yellow, blue, and some other hues, he painted small dots onto his canvas, one next to the other. He did not mix the colors, nor stroke them on. As he moved from green sections (where he used blue dots and yellow dots), to a purple section, (using blue dots and red dots), he varied primary colors the eye could see and mix according to their number and placement. Within a short distance from the artwork of course, the eye mixes blue and yellow into green, red and blue into purple, and so on.

**Texture:** The detailed application of the painted dots sets up a uniform surface texture covering the whole composition.

**Shape:** Shapes are most closely related to what the eye observes. The shapes are closed, meaning complete, and identifiable. Sometimes the people are full length, sometimes three-quarter length. Usually human subjects are in profile or turned to a three-quarter view. Volumes and mass are revealed through the rendering of natural lighting effects.

**Space:** Space is naturalistic with consistent recession according to linear perspective principles. The illustration shows a young woman powdering her face inside a room. The space used here is a simple room interior.

Seurat used oils on canvas for this painting, but the student studying Seurat needed to transfer his approach to drawing media. Most students choose oil-base crayons and relatively smooth paper when working with the pointillist process. They do preliminary sketches, complete the sketch on the paper, then begin drawing with dots of color.

***Figure 6.1*** *Georges Seurat (1859–1891).* Young Woman Powdering Herself (Madeleine Knobloch). *Oil on canvas, 1889–90. Courtauld Institute Galleries, London. Photograph, courtesy The Home House Trustees.*

Usually because of a limited studio area and compressed time frame, no one using Seurat's theory works on his huge scale. Even using an 18″ × 24″ format, a completed drawing can take several weeks.

Cubism was one of the most important theories in the history of contemporary art. Initiated by Spanish artist Pablo Picasso, 1881–1973, and by the French painter Georges Braque, 1882–1963, cubism unlocked five hundred years of verisimilitude and opened twentieth century art to virtual tides of subjective expression. Because cubism is pivotal to many processes following its time, many students opt to study cubism.

The cubist theory developed somewhat in this manner. Linear perspective gave us a fixed field of vision from one station point. Our images were stable and appeared physically real. But without a fixed field of vision, we must assume many points of view, seeing all sides of an object. However, our eye focuses only one place at a time. An image built from a simultaneous multiview fractures the natural object but never quite loses its recognizable nature. In other words, a head always remains a head. Overlapping voids and solids create a new geometry that transforms the

**Figure 6.2** *Cathy Shannon. A figure drawing according to Seurat. 18″ × 24″, oil based crayons on rag paper.*

**Figure 6.3** *Pablo Picasso (Spanish, 1881–1973). Daniel-Henry Kahnweiler. 1910. Oil on canvas. 100.6 × 72.8 cm. Gift of Mrs. Gilbert W. Chapman in memory of Charles B. Goodspeed, 1948:561. Photograph © 1991, The Art Institute of Chicago. All Rights Reserved.*

natural appearance, *what we see* from one point of view, into an abstract realism, *what we know* about the object from all sides. That image finds a new space: a vertical, sculptural, bas-relief-like facade.

Picasso's portrait of *Daniel-Henry Kahnweiler* displays the new analytical realism. We notice right away that the figure is not lit with natural light, nor does it have atmosphere. The Kahnweiler portrait incorporates many of the organizing principles of analytical cubism (identified by art historians H. H. Arnason, H. Gardner, and H. W. Janson):

1. Excludes the representation of light and atmosphere
2. Uses a narrow range of colors, almost monochrome
3. Abandons movement, naturalism, and linear perspective
4. Views subjects from all sides (depicted objects as they are known, not as they are seen at one particular moment) reconstructing an image with geometric units, sometimes overlapping
5. Maintains a shallow picture plane, a bas-relief-like facade
6. Sustains nature as a conceptual realism, not a visual realism.

Bearing the list in mind, study the examples and consider the information before initiating your cubist drawing.

**Figure 6.4** *Kevin Scott. A cubist drawing. 12″ × 15″. Ink and rag paper.*

**Figure 6.5** Gene Rauch. A cubist drawing. Ink, colored papers on rag paper.

**Figure 6.6** Lee Ann Takes. A cubist drawing. Graphite on paper.

**Figure 6.7** Li Trout. A cubist drawing. Graphite on paper.

**Figure 6.8** Linda Dainty. A cubist drawing. Graphite on paper.

**Figure 6.9** *Doug Myers. A cubist drawing. Charcoal on charcoal paper.*

You have seen six student drawings that employ the cubist theory, each drawing quite singular. All students incorporated the principles of the theory. By now you understand that no matter how many artists might draw *the same object*, each drawing will reflect the uniqueness of the person drawing. So, too, with theories. The object of any study is therefore, not to set out to be different. Define the main points of a theory, draw with the theory in mind, and your drawing will be distinct.

Beginning to draw with cubism in mind is difficult for some. A few suggestions here. Let's say first what cubism is not.

Cubism is *not* just flat patterns and designs. Straightening eyebrows, lips, mouths, and ears simply reduces a head to a Halloween pumpkin face. Each shape, meaning small sections, like cubic units, of the head, face, neck,

whatever, needs to be observed and drawn from several points of view. Reconstruction of a cubist head means placing those cubic units together to embody the fractured view.

Shapes are *not* just features. Try looking at repetitions of shapes. For instance, the eyebrow, the curve of the top of the ear, the upper lip, the upper eyelid, the nostril—are there ways to incorporate portions of one of those units with another to create a new multiviewed shape?

Cubism is *not* a series of profiles turning in space. Remember cubism does not work with motion, nor with full profiles.

Cubism is *not* a series of random features separated and floating on a piece of paper. Loose eyebrows, lips, and noses scattered over a field of white paper is not analyzing cubic shapes from many viewpoints and reassembling them.

Cubism is *not* surrealism nor contemporary video illustrations. Turning your model's head into a skull inside a melting clock is not this theory either.

I would suggest you start with a small drawing, say a 9″ × 12″ paper. Pose the model so that you can move all the way around him or her, looking at the head. Take one position, draw a small section, not features, but perhaps a portion of the nose and lip. Move to another location. Look at the shapes, the cubic units, available to you there. Are there some of those units you can coordinate with the portion of the nose and lip you just drew to begin to construct a new multiviewed cubist head? If you continue analyzing shapes from a number of points of view, you can teach yourself how to reconstruct the head or figure before you.

Often students take several days to work their way into this theory. It is one of the more difficult ones to understand.

After looking at some illustrations and reading what cubism is and is not, you can reinforce your learning by returning to the student examples once again. See how each student chose to rebuild heads, body, and backgrounds. Take special note this time, while looking at the examples, of the use of lighting or modeling or lack thereof.

Hereafter you will need to identify for yourself how each artist used the formal elements: line, value, color, texture, shape, and space. Theories will be defined generally, but part of the learning process is chasing the mystery.

Marcel Duchamp, 1887–1968, another Frenchman, moved motion into cubism. Duchamp was influenced both by Marey's chronophotographs—which showed in one photograph a moving figure in a series of stop actions—and by Edweard Muybridge's continuous panels of figures in motion.

Within several paintings around 1911–1912, Duchamp infused cubism with movement, most notably one titled *Nude Descending a Staircase, No. 2*. However, he kept the restrained tonalities of the earlier theory. He remained analytical with the figure and its surroundings, but he emphasized simultaneous successions of the moving, motive, or energetic body lines as the figure descended a staircase. He also placed the viewer in the center of the picture.

**Figure 6.11** *John Williams. A figure drawing according to Duchamp. Pastels on charcoal paper.*

**Figure 6.10** *Marcel Duchamp (1887–1968).* Nude Descending a Staircase, No. 2. *1912, Oil on canvas, 58″ × 35″. The Philadelphia Museum of Art. Louise and Walter Arensberg Collection.*

Our student obviously has studied and understood cubism in order to move into Duchamp's theory of cubism with motion. But the student varied his subject by asking the model to move *up* the steps not down.

Franz Marc, 1880–1916, a German, developed his work by venerating nature. Much of the basis of his art was principled in cubism, but he selected other devices of distortion, and he used brilliant colors. Among other historians, H. H. Arnason has observed Marc's use of color was less descriptive than symbolic; blue as masculine, yellow as feminine and red as matter. Marc's color shapes took on less graphic configurations than they did spiritual or mystical. In other words, figures or animals were not drawn as such but were used to springboard into abstract shapes similar to the rectangular geometry one sees in brilliant,

translucent Gothic stained glass windows that suggest spiritual or mystical experiences. And, Franz Marc's work generally was parallel to the picture plane.

Our student used the model in the classroom but drew many sketches from nature to analyze and recompose her work in line with Marc's approach. Her drawing is small, about 9″ × 12″, with densely brilliant reds, yellows, blues, oranges, violets, and greens. Her medium was colored pencil.

Max Beckmann, 1884–1950, a German, worked less with pictorial truths than with very large, complicated allegorical compositions within the delimited space already familiar in cubism. Very often Beckmann's subjects dealt with suffering (in one work one figure is carrying another upside down, to represent the dragging of one's own memories and failings alongside). But throughout his work are brilliant portraits of individuals. Repeatedly one finds the sitter's skin tones painted in muted colors overlaid with whites, a simultaneous sense of living and dying. Sometimes proportioning is close to natural, more often it is not. Body attitude, limbs, and posture reference the distortions reminiscent of cubism. Settings also assert the cubist delimitation of space.

**Figure 6.12**   *Franz Marc (1880–1916). Stables. 1913–1914. Oil on canvas, 29 1/8″ × 64 1/4″. The Solomon R. Guggenheim Museum, New York.*

**Figure 6.13**   *Tracy Hilton. A figure drawing according to Franz Marc. 9″ × 12″, colored pencil.*

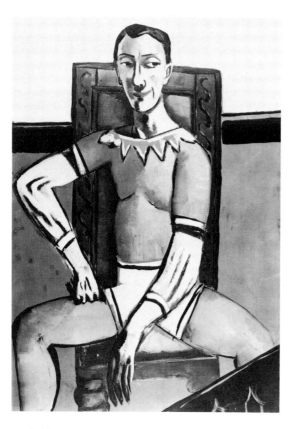

**Figure 6.14** *Max Beckmann (1884–1950).* Portrait of Zeretelli. *The Dancer. 1927. Oil on canvas; 55¼″ × 37¾″. Courtesy of the Fogg Art. Harvard University, Cambridge, Massachusetts. Gift of Mr. and Mrs. Joseph Pulitzer, Jr.*

**Figure 6.15** *Paul Guy. A figure drawing according to Max Beckmann. Pastels on charcoal paper, 24″ × 18″.*

*Portrait of N. M. Zeretelli,* is just such a work. The back wall is parallel to the picture plane. The chair has a disjointed perspective. The figure retains mass in the face and torso, flattens in the upper right shoulder and the boxer shorts but invites the illusion of volumes in the rest of the body. Lines are heavy and dark. The portrait is austere.

Using Beckmann's pictorial language in the next illustration, a student work, we see similar distortions in the anorexic body shapes, the dark, heavy lines, the purple, blue, grey, and white colors, and in the dramatic values. Two figures are in this work. The student's two-point perspective was not common with Beckmann, although he did use it. Usually Beckmann's figures are seen straight to the picture plane, while he distorts surrounding space, both interiors or exteriors.

We turn now to an ingenuous Frenchman, Henri Rousseau, 1844–1910, an unsophisticated, untrained artist who ranks among great artists as a primitive. He generally placed figures, animals, jungles, or deserts parallel to the picture plane, flatly. Silhouettes are prominent in his work, as are low key colors. Because small plants or leaves were transformed into a larger scale for trees and tropical scenes, exquisitely rendered with eerie lighting, mystery pervades

many works. Space is much like a stage set with wing flats and border pieces; that is, thin, vertical, movable sections of painted scenery.

*The Dream* is such a painting. A luminous nude reclines on a Victorian couch in the middle of a jungle with birds and elephants and lions, oh my! In this instance the scenes are occasionally dotted with strange people.

The student who adopted Rousseau's primitivism sketched from the model in the classroom after posing her in a prone position on the model stand. Outside of class he found leaves, grasses, and small branches across campus. At a botanical center he sketched different types and kinds of leaf patterns and growth. Taking his sketches and collected items with him, he concentrated on creating his delightful Rousseauian composition, bursting the bounds of space by exceeding the picture plane edges, not unlike what Rousseau implied in his works.

Marc Chagall, 1889–1985, a Russian, spent much of his life painting reminiscences of his childhood, his marriage, and his faith. His approach was much less a theory than a method. He did not use visible reality. He allowed images to surface in his work from within, permitting a painting to materialize. He worked in poetic, visionary,

**Figure 6.16** *Henri Rousseau (1844–1910). The Dream. 1910. Oil on canvas, 6'8 1/2'' × 9'9 1/2''. Collection, The Museum of Modern Art, New York. Gift of Nelson A. Rockefeller.*

**Figure 6.17** *Adrian Penn. A figure drawing according to Henri Rousseau. 18'' × 24''. Colored pencil on smooth rag paper.*

*Figure 6.18* *Marc Chagall (1889–1985).* Equestrienne. *1927. Gouache on paper, 20 1/8" × 26". Private collection, Switzerland [Seen in* Marc Chagall *by Werner Haftmann. Translated by Heinrich Baumann and Alexis Brown, Harry N. Abrams, Inc., New York. Colorplate 24, page 112].*

symbolic scenes. His is a world of fantasy—of a woman lying in moonlight atop a mythical horse, of the sense of the love of a man for a woman as he floats above her, of big birds and small cows, of red roofs and music and poetry.

Chagall's space sometimes resembles mystical cubism, sometimes not. Almost always there is an underlying theme of joy. Discontinuous shapes are devised, allowed. Colors are vibrant, brilliant.

The student who adopted Chagall's methods studied many of his works, adapted dream fantasies of her own using Chagall's space, cityscapes, moon, and animal shapes. Hers is a large work, pastels on black paper.

Within the third chapter we touched on Surrealism, a movement that sought to use another form of realism, that of the subconscious and the dream. The surrealists wanted a super reality. You remember that surrealist artists were on the frontiers of expanding the matter-of-fact, the physical presentation. They wanted to move away from logic to impulse with no moral or aesthetic boundaries. They wanted logical disconnection. They incorporated loss of function. They concentrated on the fantastic and the grotesque.

*Figure 6.19* *Ann Johnson. A figure drawing according to Marc Chagall. 4' × 6', pastels on black paper.*

We will look at two surrealist artists whom students have chosen to study. The first is Salvador Dali, 1904–1989, a Spanish artist whose academic polish of the realistic people and objects in his works made their displacements all the more horrific. Even the title of his work, *Dream Caused by the Flight of a Bee around a Pomegranate a Second before Awakening*, raises one's level of alertness.

Generally, Dali placed things in the foreground before deep, often sandy, desert-like distances. Within the dramatic recession of space he juxtaposed placement of other objects. For instance, in this work, are the lions leaping over the figure or onto the figure? Where did they come from? Is the gun drawn vertically, diagonally, or does it originate on the same plane as the horizon? Whoever saw an elephant with spider legs? And on what is the nude resting, or is she slightly suspended?

The student's application of Salvador Dali's surrealism includes the use of his space, the spider legs, the disconnected, beautifully rendered realistic images that give us a surrealistic dreamscape.

Rene Magritte, 1898–1967, a Belgian, was another surrealist, but his work was not based on the "let-anything-surface-from-the-unconscious" school of autonomic surrealism. Rather, Magritte wanted to reveal alternative meanings of people and things placed in different environments. His work juxtaposes objects and his images startle. He introduces different functions for some things, such as a bleeding statue. Always there is intelligence and wit in his works.

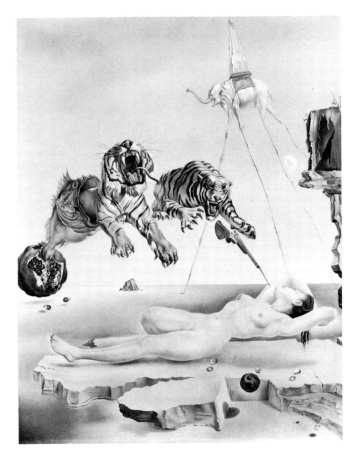

**Figure 6.20** *Salvador Dali (1904–1989).* Dream Caused by the Flight of a Bee around a Pomegranate a Second before Awakening. *1944. Oil on wooden panel, 51 × 41 cm. Thyssen-Bornemisza Collection. Lugano/Switzerland.*

**Figure 6.21** *John Dorn. A Surrealist drawing according to Salvador Dali. Graphite and pastel on charcoal paper.*

**Figure 6.23** *Rene Magritte (1898–1967).* The Fair Captive, II. *1935. Oil, 18″ × 25 1/2″. Collection, Robert Strauss, London. [From* Rene Magritte Catalogue *by James Thrall Soby,* The Museum of Modern Art, New York, *Distributed by Doubleday and Company, illustration page 35]*

Our student used both of these devices, drawing three sections of his nude, each framed, placing her on the front, or a little behind ( we are not sure) the first canvas. Between the two canvases we see a vertical portion of the same nude. Behind her on another canvas a third image, a torso, is drawn surrounded by a large drawn framing device. Finally, the artist placed the drawn canvas units into the middle of a field which is both a part of the drawing and apart from it.

Our last work illustrates Pop Art, a movement based on mass-produced urban culture as seen in commercial products, movies, comics, and magazines. Rejecting any distinctions between good and bad taste, the pop artists excerpted the most brazen parts of our popular culture and elevated them to a fine art. They used soup cans, electric chairs, sex, and highway signs as subject matter. The works often incorporated hard edges, bright, flat billboard colors, and the printer's ben-day dots. The illusion of space, if an artist used it, was likely to be carried by line rather than value.

Tom Wesselman, 1931–, an American, is a pop artist. He is an artist of the TV commercial and movie magazine sex. Many times he made assemblages incorporating actual objects like clocks, TV sets, and such. He painted a quantity of "Great American Nudes," numbering them as he went. Wesselman's nudes are nonsensual, consistent with the dehumanizing numbering system in his titles. His works illustrate the vacuousness of the American sex symbol.

The illustration below is a drawing created from a nude model using Wesselman's pop-art approach. One of the decisions this artist made was to change both the width and colors of the stripes on the blanket to reveal more closely the simplistic presentation of Wesselman.▲

**Figure 6.22** *Rene Magritte (1898–1967).* The Eternal Evidence. *1930. Oil on canvas; five panels, top to bottom, 8 1/4″ × 4 3/4″, 7 5/8″ × 9 1/2″, 10 3/4″ × 7 5/8″, 8 3/4 × 6 3/8″, 8 3/4 × 4 3/4″. William N. Copley Collection. [From* Rene Magritte Catalogue *by James Thrall Soby,* The Museum of Modern Art, New York. *Distributed by Doubleday and Company, illustration page 32]*

We show two pieces here because the student who adopted this theory incorporated like-minded items in his work. *The Eternal Evidence* displays a segmented nude in five units. *The Fair Captive, II,* shows an easel on which an artist's canvas rests. The image on the canvas is coincidental to the field in which the easel stands. The painting is a play on spatial illusion.

**Figure 6.24**  *Alan Leusink. A surrealistic drawing according to Rene Magritte. 18″ × 24″, graphite and colored pencils.*

**Figure 6.25**  *Tom Wesselman.* Great American Nude, No. 57. *1964. Collage, 48″ × 60″. Private Collection [Seen in* History of Modern Art *by H. H. Arnason, Prentice-Hall, Inc., Englewood Cliffs, N.J. and Harry N. Abrams, Inc., New York. Illustration 1045, page 583]*

---

**113**

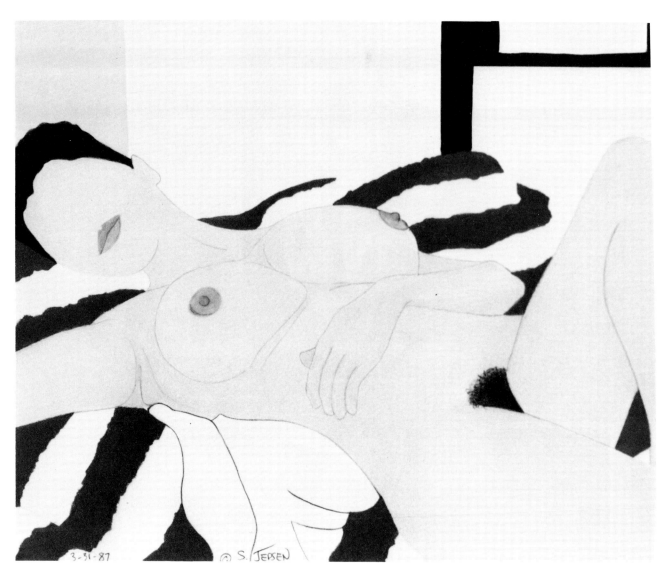

**Figure 6.26** *Scott Jepsen. A Pop Art drawing according to Tom Wesselman. Graphite and colored pencils on rag paper. 12″ × 18″.*

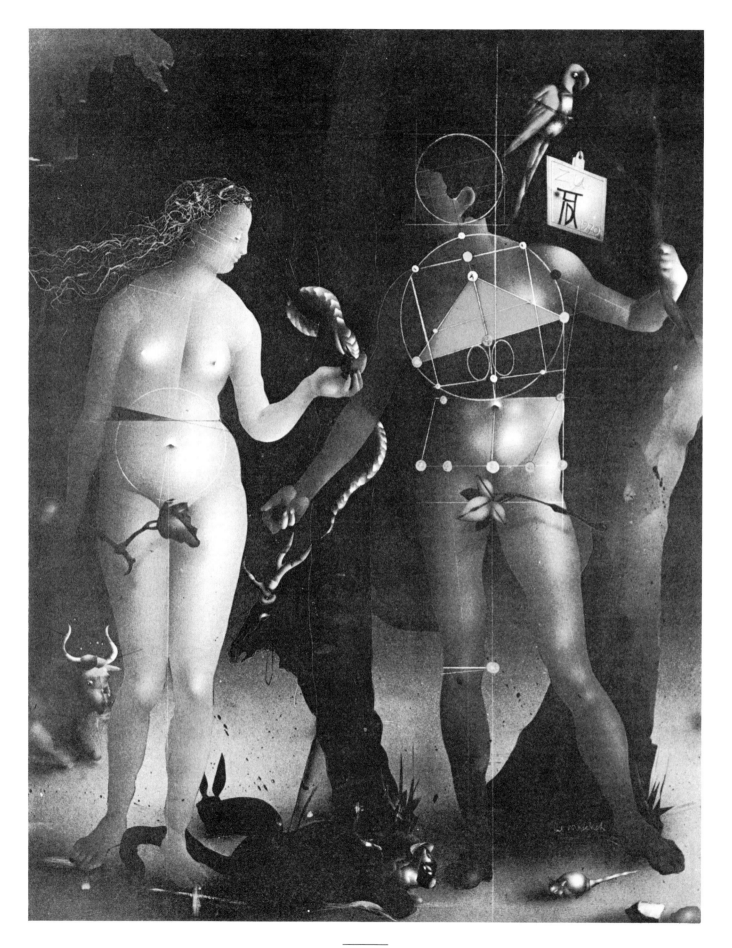

# Two

## The Expressive Paradigms

# Introduction: The Expressive Paradigms

This book began with a simple insight that took years of studio teaching to discover. There is no neutral way to draw the figure. Every time the task is attempted, whether through complex anatomical exercises or simple contour practice, something expressive starts to happen. A personal imprint. It seems we make an unspoken commitment to our feelings whenever we apply our strokes toward a rendering of the nude body. Typically, studio teaching seeks to discipline the process through methods that encourage students to reveal on paper something that approximates what they see before them. What results may be pleasing, often because the image conveys *a shock of reality,* something a viewer could almost *reach out and touch.* The good student appreciates that he or she has learned one approach to the creation of an illusion, an illusion dear to Classical Greece and the Renaissance, and dear to a great many art academies and commercial illustrators. The student's mastery of representation also extends an expressive opportunity. That opportunity is what the following expressive paradigms are about.

Picasso has been quoted as saying, "I used to draw like Raphael, but it has taken me a whole lifetime to draw like a child." In a sense Picasso was saying that it had taken him years to draw like himself. Through the paradigms, you will be touching personal roots within historical roots. Roots make things grow, make things mature. Each paradigm stands alone as a venue to expression, and investigation may begin with any of them, and proceed in random order. Chances are, you have already started with the first paradigm, *Figure As Form,* by undertaking the prior exercises of this book. What we are now after are springboards to broaden and deepen figurative issues in your journey through your stroke and through your singular response to the human body.

Why the human body? Put simply, it's where you live. It's the only true residence you will ever know. Its features are the most intimate physical definitions of your self-image, and that of your subjects. Growing, changing, aging, dying, being adorned or unadorned, dancing or fighting or posturing, everything the body does is rich with clues to our nature. Furthermore, the body scales our world. Things in the environment are big or small, balanced or unbalanced, symmetrical or asymmetrical, threatening or friendly, according to human measure.

Some of the historical paradigms will seem more analytical than expressive. Parts of *Figure As Form* will correlate notions of ideal beauty with proportions that can be measured in rather technical ways. Ideal beauty gives way to a freer exploration of passions in *Figure Against Form,* but requires that students understand the forms artists used as working assumptions. *Figure Above Form* demands an archetypal awareness of the human body as set within contexts that bring to it the power of the icon, religious and cultural. The accidental and surreal, anti-art and ambiguity pervade *Figure and Form, a Paradox.* Finally, *Figure, the Fullfiller of Form* draws on everything we know about narrative, and about the storytelling continuities contained in our imaginations and histories.

But not all is analytical. Within the paradigms are suggestions for experimentation. Called *Studies,* these suggestions, or descriptions of someone else's process, are intended as stretching exercises for your stroke, for the unique way you bring information to problems generations of artists have struggled with. Just remember, you talk to your stroke, and it talks to you. You will be both inside and outside of your venture at every step. You will be choosing, acting, reacting. If you started at the beginning of this book, you have already moved through a number of exercises. The question is, what have you been "forming" through the use of the figure?

But let's pause for a minute and review the key words and terms that will come up frequently in what follows. *Form* is something you make, a construct *and* an image. Form has a beginning and an end, that is, it has qualities that define its shape, scale, and completeness. That's because the completed picture and its physical shape are made through choices. Those aesthetic choices derive from your emotions, your intellect, and the physical process of forming. The *figure* provides your *subject matter*. You, the artist, supply *expressive content*. The work, not the artist, supplies an *affective meaning* for a viewer, whatever a viewer is enabled to feel and think. In the act of *forming* you select your materials, your media, and your scale. You have your way of making strokes, your methods, your style. With instinct or deliberation you choose how to use the *formal elements* most necessary: line, value, color, texture, shape, and space. Through decisions you make in creating a visual composition that uses the figure as subject matter, you eventually arrive at your picture. You have an end result, a construct. But now you need to ask yourself, What has been formed? What has happened in reaching for a *correspondence* between the subject matter and the image? In other words, what is being expressed?

Often this question is best asked after the work is complete. In fact, the process of answering the question is part of what tells you the work *is* complete, and has its form. Here's where you release the work to its independent life, and you relate yourself to its affective meaning independently.

Join me now in looking at some of the oldest ways of using the figure: the initial paradigm, where the figure is the form and the form is the figure. That correspondence is among the greatest of the legacies we have from the Italian workshops and studios of the fourteenth and fifteenth centuries. But their heritage went back even farther, and their impact still animates whatever we might view as contemporary.

Paradigm I

# Figure as Form

▼

**F**igure as form means simply that the reproduced figure is in itself a complete expression, provided that the reproduction derives from an ordering concept. To understand this we may do well to reflect for a moment on our visual legacy. As you reflect, remember you are seeking something more catalytic than analytic. Personal expression is the objective, but recovering some of the living belief and power passed on by each generation in the history of western art can be shared.

Consider first two semi-footnotes. For much of the history of art the word *artist* was not understood in the self-conscious, largely romantic way we have associated over the past century or so. From oldest times an artist was primarily a craftsperson, a skilled worker, often a member of a guild whose role was the dispensing of images for a purpose (and usually for a commission). Expression was more or less the unconscious signature that resulted whenever the craftsperson

1. embodied in his/her work something of what was understood to be beautiful or compelling in nature (the world "out there"), or
2. embodied in his/her work qualities of character, usually heroic or honorable ones, that exemplified the moral weight of a subject.

Typically, the former method required an image of the unmovable moment (perhaps in dance, combat, romance, or contemplation, etc.) that would otherwise be lost to the temporality of life. The latter method, often seen in portraits, served to honor and to instruct, and to make art a fruitful extension of philosophy, engendering ultimately what we know as *aesthetics*, the science of the beautiful. All such concerns were meant to elevate the social context in which they existed. And all required some kind of opera-

tive ideal: *what ought to be seen (a person's finest hour, for instance)* were it not otherwise hidden by fashion, ignorance, accident, or corruption.

Few of today's artists struggle with the question of how beauty is perceived and known, or toward what magnificent social end is the act of doing art an ennoblement. Even the concept of an *ideal* seems alien to the modern mind. But for the longest history of art such a concept was *the* expressive touchstone. And when artists/craftspersons strove to render the figure on panels or in stone, they often saw their subject matter in the terms their religion taught them: the image of the deity. god-like. The human figure, the one we all inhabit, was the surest clue to ideal beauty, and to all subsequent reflections of how organic, three-dimensional parts could best relate to a whole in terms of scale, symmetry, and proportion.

The ideals of beauty and character are still productive premises if we approach them as imperatives toward the highest subliminal vision we all carry. By seeking and honoring them, we steer clear of the cute and pretty, or the trivial and banal. And much as earlier artists understood that beauty was something to be wrested or liberated from the prison of the blank slate or the stone block, so did they free within themselves the hidden creative powers they believed were bestowed in their nature. In this sense, rendering the human figure as a mirror of an ideal was an exercise toward maturity and selfhood, indeed the best possible visual exercise.

Of course, human subjects come in all shapes, sizes, ages, and demeanors. Which ones approximate an ideal? Here's where the artist/craftsperson had to search deep inside. What were his/her profoundest notions of the beautiful? (It's the same search we all make.) Many of these skilled workers found that a correlation of perception and imagination could derive from an application of purely formal abstractions: numbers. Ancient philosophy had championed the elegance of numbers and geometrics. Surely, as wisdom pointed toward an ideal; it also provided tools, something all craftspersons appreciate.

As you work with the tools described in this chapter, you are testing your personal sense of the way an image of the human figure in and of itself satisfies requirements for a complete artistic statement. Figure *as* form. Form *as* figure. Undoubtedly, you will move ahead toward new tools that will expand the legacy. Artists do use systems other than mathematics. And sometimes a system only pertains to the structural application of materials, media, or methods. Perhaps you have seen small photographs blown up. Adjusting the scale of an 8″ × 10″ photograph to an 8′ × 10′ drawing requires a system, a dependable, replicatible method.

Between the Egyptians of c. 2400 B.C.E., and Albrecht Durer of c. 1530 C.E., most systems used numbers to point the way to ideal beauty. Then, after a 500-year hiatus, during which pursuit of an expressive ideal turned from numbers, Corbusier, a French architect in the 1930s and 1940s, reverted to the Golden Ratio of ancient Greece, and to a proportion based on the male figure. Corbusier sought to secure consistency in numerical diagrams that would allow fellow architects the world over to create buildings by using his mathematical vocabulary much the way a musical composer uses notation. And to honor human scale.

Typically, mathematical systems were used to construct symmetrical, balanced figures. Today the use of basic geometric forms with precise ratios for body parts has fallen largely into disuse in the fine arts. So why study *systems*, if few really believe in them anymore? Well, because we do believe in them, admittedly or not.

Consider this. How many times have you looked in the mirror and witnessed the asymmetry of your own face—how displeasingly the mouth works, how crooked the teeth, how lopsided the features, droopy, pointed, flat, or whatever? If you have done this, you are secretly asking for more order, more harmony, better proportioning, something closer to an ideal.

Art historian Rudolf Witkower says this about the importance of your mirrored perception: "Symmetry, as the balance of parts between themselves and the whole, is a primary aspect of proportion. Bilateral symmetry is only one of seventeen species of symmetry. It is the symmetry of the human body, and for that reason of towering importance to mankind . . . Every disturbance of the balance of parts (e.g., a short leg, a crippled hand) evokes reactions such as pity, irritation, or repulsion. When all is said and done, it must be agreed that the quest for symmetry, balance and proportional relationships lies deep in human nature."[1]

**Figure I.1** *John Baldessari*. Repair/Retouch Series: An Allegory About Wholeness (Plate and Man with Crutches). *1976. [Seen in* Art in America, *May 1981, page 133.]*

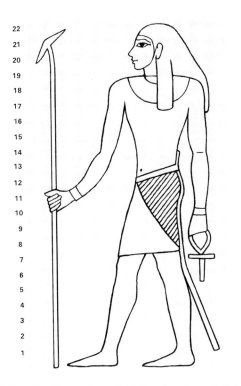

**Figure I.2** *The "Later Canon" of Egyptian Art-c. 2400 B.C. [after Travaux relatifs a la philologie et archeologie egyptiennes, IIVII, 905, p. 144—seen in* Meaning in the Visual Arts *by Erwin Panofsky, The Overlook Press, Woodstock, N.Y., 1974.]*

**Figure I.3** *Christine Thelen.* Tomb of the Understood. *1990. Mixed Media. Mummy, 74″ × 28″ × 8 1/2″. Sarcophagus, 76″ × 30″ × 10″.*

Egyptians of the third millennia B.C.E. used the square. Because the square is absolutely as high as it is wide, it is a deceptively simple shape. Yet, we can derive root rectangles from it, the Golden Ratio, even the logarithmic spiral, which is known to be in plants. Sunflower centers, for example.

The premise behind the Egyptian system was the fixing of an identity for eternity, a container for the eternal spirit. Strict mathematical rules applied so that any deceased Egyptian correctly reproduced might, by an appropriate ritual, be called back into being.

The Egyptian artist/craftsperson used the same graphing system for all figures.

He knew " . . . from the outset that he must place the ankle on the first horizontal line, the knee on the sixth, the shoulder on the sixteenth, and so on."[2]

## A Study:

Through a protracted but very interesting process, Christine Thelen mummified her body and placed the sculptured shape within a sarcophagus on which she drew images according to the Egyptian Canon.

The process went something like this: Thelen began by dividing her body into three units (legs, torso, neck/head) for making the mummy. Sitting on the floor, legs extended, she covered the top portions (left side to right side) of her feet, legs, and thighs with lightweight plastic sheeting over which she draped wet plastered gauze strips.

When she began the torso, this time lying on the floor, help from a friend was necessary. A third *sitting* was needed to work with the head. She and her assistant kept air passages free, as the plastered gauze strips were wrapped over her plastic-covered head.

Each of the three body units took over one hour for wrapping, then holding still for the gauze to dry. Once the three large pieces were removed and dried, Thelen spliced them together with gauze. To keep the mummy stiff, polyester batting was stuffed inside the half-shell of the gauzed body form.

The second unit to be built was the sarcophagus, which was made from sturdy muslin stiffened with white glue. Thelen chose chalk pastel, watercolor pencils, and markers for the cover, which was divided into five panels, top to bottom. Colors were chosen to "represent lapis lazuli . . . the other colors are typical of Egyptian wall paintings."[3]

Thelen followed the rigorous standards of Egyptian tomb art in several ways. The funerary portrait depicted the person in her youth. The pattern around the head and at the feet was taken from a death robe laid over a mummy before being buried. The second panel indicates both a neck piece and guardian figures who protect the dead in the afterworld. The center panel figures represent Thelen's mother and father on either side of a lotus column, a symbol for life. The seated male and female figures in the fourth panel represent her sister and brother who face an adult figure, symbol of instruction.

**Thelen learned that Egyptian hieroglyphs were based on sounds. She modified those symbols for ones of her own and wrote the history of her family on the sarcophagus.**

In following such rigid schemes, the Egyptians overlooked three very important things about drawing figures. They ignored movement of the parts, they ignored foreshortening, and they ignored the viewer's position.[4]

These shortcomings were effectively overcome by the Greeks. But it took nearly 2,000 years to do so. In the fifth century B.C.E., Greek artists allowed for organic movement, foreshortening, and corrected optical illusions. But their substratum data for explaining all visible phenomena were *numbers*.[5] They used numbers to make their figures more perfect, more beautiful, more ideal.

For working artists, *ideal* can have two meanings. One way of idealizing has the artist working with natural appearances but improving on them, straightening the nose, widening the eyes, and so on, following a numerical system for doing this *improving*. A second way was articulated by Plato. Plato said that anything we see is an imperfect copy of an unchanging and imperceptible Idea or Form of that object. As artists, our visual systems for achieving this philosophical posture are important, even though they always fall short, the *finally perfect figure* is in the mind and always remains an ideal. The artist/craftsperson begins not from nature but from intuition.

The Greek artist began with measurements directly from the human figure, not like the Egyptians, who designed their figures from a network of squares. The Greek sculptor is said to have started with the smallest part of the body that could yield a common fraction. That fraction might have been the tip of the little finger. Using that module, say, 3/4″ from fingertip to first knuckle, the artist would create the height and width of a figure with multiples of that module. An arm for instance, wrist to elbow, might be 14 of those unit lengths. (Remember, our twelve inches, one foot, had as many as 280 variants in Europe as late as the 18th century, often taken from the length of a king's foot.) The Greek artist would continue throughout with the *tip-of-the-finger* module. The distance from the far left to the far right eyebrow might be five modules, and so on. The end result in the drawing meant that the model's body was elongated or shortened according to the *module system* the artist used, making the body and its features more symmetrical, unified and perfect. Once the figure was completed, the artist used his judgement for small corrections here and there, now relying on his own intuition. The system alone did not create *the beautiful*. But the numerical system was the basis for much figurative work by Greek artists.

Let's look first at a Roman copy (most information we have about Greek statuary comes from Roman copies) of the *Spear Bearer*, by Polyclitus.

***Figure I.4*** *Polyclitus.* Doryphorous (Spear Bearer). *Roman copy after an original, c. 450–440 B.C. Marble, height 6'6″. National Museum, Naples. Art Resource, NY.*

Now look at the drawing of the *Spear Bearer*, with one arm flung over his head. The body stance alone, regardless of the proportions, skewers the work. The *symmetria*, the harmony of parts, is lost.

To draw an ideal figure through *many numbers*, we need to understand proportion. Proportion does have to do with mathematics and geometry. (Those interested in pursuing the use of numbers should refer to the footnotes for this chapter and to the simplified explanations of geometric proportion, arithmetic proportion, and harmonic proportion. Included also is a discussion of how to construct the Golden Rectangle and the Root Five Rectangle.[6] The Golden and the Root Five Rectangles were used for determining ratios of figures, for selecting dimensions of drawing papers, and for applying the mathematics of architecture. Both types of rectangles were most important to artists and architects up through the Italian Renaissance, about 1550 A.D.)

## A Study:

**Richard Shook began with the square and created a Root 5 Rectangle. Once that proportion was secure he *adjusted* the model's body proportions to fit the square, all the while allowing the *scaling and proportioning evidence* to remain visible within the drawing. Expressively, this drawing becomes a playful way of drawing "on" the picture plane with geometric shapes, as well as "in" the space that contained the figure behind the picture plane.**

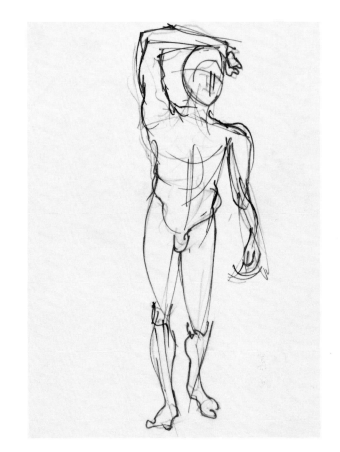

***Figure I.5*** *Drawing of the spear bearer with arm and head askew.*

***Figure I.6*** *Richard Shook.* Drawing Using the Root 5 Rectangle as a Format. *Graphite on rag paper, 6″ × 9″.*

Several hundred years after the Spearbearer, Pollio Vitruvius, a Roman architect active circa 27 B.C., wrote on proportions and symmetry taken from ratios in a human body but used in temples. The *well-shaped man* Vitruvius saw as a perfected figure, principled on the following proportions, some of which are still taught today as rule-of-thumb guidelines.

Of the total length, 6′ male:

1. Head—crown to chin = 1/8
2. Face—hairline to chin = 1/10
3. Pit of throat to hairline = 1/6
4. Pit of throat to crown = 1/4
5. Chest width = 1/4
6. Hand—wrist to tip of third finger = 1/10
7. Length of foot = 1/6

Vitruvius goes on to say, "Then again, in the human body the central point is naturally the navel. For if a man be placed flat on his back, with his hands and feet extended, and a pair of compasses centered at this navel, the fingers and toes of his two hands and feet will touch the circumference of a circle described therefrom. And just as the human body yields a circular outline, so too a square figure may be found from it. For if we measure the distance from the soles of the feet to the top of the head, and then apply that measure to the outstretched arms, the breadth will be found to be the same as the height . . ."[7] He is describing plane surfaces, which are perfectly square.

You will remember that the circle and the square are perfect geometrical forms, now imposed on a *perfect* figure.

Cesariano, editor of the *Como Vitruvius*, placed the square inside the circle. The man appears ape-like in his proportions and bears a small head.

**Figure I.7**  *A head drawn to Vitruvian Measurements.*

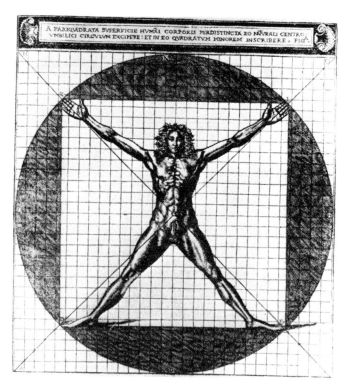

**Figure I.8**  *Cesariano.* Vitruvian Man. *From the Como Vitruvius, 1521. [pg. 38,* The Nude *by Kenneth Clark. Doubleday Anchor, NY, 1956.]*

Leonardo da Vinci, a close contemporary of Cesariano, about 1450 years after Vitruvius, understood the Vitruvian Canon this way: to have the navel as a center, the figure must spread his legs only far enough to form an equilateral triangle, and spread and lift his arms only as far as the crown of his head—then one can contain the figure in the circle and the square.[8]

As the Roman Empire waned, the naturalistic and ideal imagery from Greece was abandoned. Early Medieval Art, circa 300–1500 A.D., reverted to the use of planes, but not as the Egyptians used planes. The methods used in the Middle Ages were schematic, utilizing one of two procedural diagrams, the *Byzantine* or the *Gothic*. The Byzantine tradition was closer to the Greek Classical tradition in that it did recognize relative proportions in nature. But the Byzantine canon was not a codex of human proportion, rather a simplified routine of multiples. The face, now *the seat of spiritual expression*, became the unit of measurement.[9] The

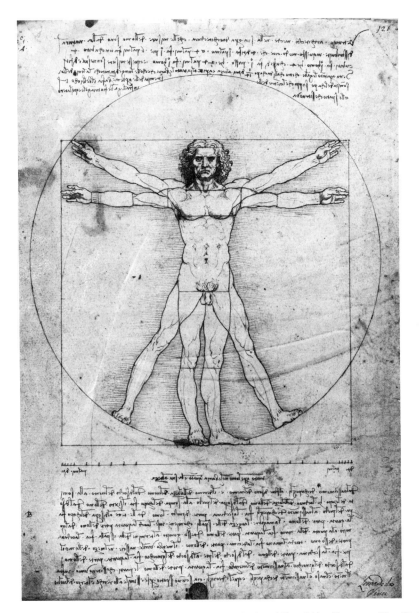

**Figure I.9**    *Leonardo da Vinci (1452–1519).* Vitruvian Man. *Academia, Venice. Alinari/Art Resource, New York.*

Greek canon, you remember, was measurable in fractions. The Byzantine Canon was based on one rather large module, the nose length. According to the *Painter's Manual of Mount Athos,* codified by Dionysius of Fourna, an 18th century monk, the figure was divided this way:

Face = 1 unit, divided into thirds, each the length of the nose.

Throat = 1/3 unit

Chest = 2 and 2/3 units wide

Torso = 3 units

Upper/Lower Leg = 2 units each

Foot height = 1/3 unit

Shoulders = 2 2/3 units

Forearm, arm, hand = 1 unit each.

A figure constructed from those measurements might look something like this scheme, nine face lengths tall.

The Byzantine method was not devised to render a *real* figure. Measurements were applied in repetitions of figures to suggest universal harmony. Perfection again was spiritual, not physical.

Using the nose length, one can see in the *schematic* and in the *Deesis Mosaic* examples that the center of the radius was at the root of the nose. Three nose lengths equaled hairline-to-chin, or the face length.

**Figure I.11** The "Three Circle Scheme" of Byzantine Art.

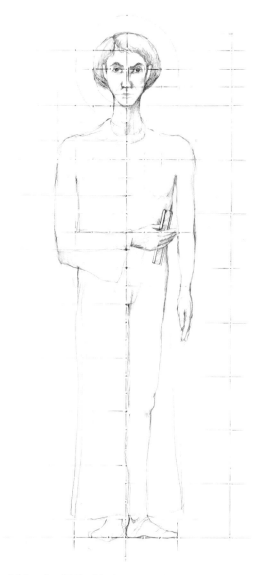

**Figure I.10** *David Pavlik.* A Man Defined by Byzantine Units of Measurement.

**Figure I.12** Byzantine Christ. *(c. 13th century.) Detail of the Deesis Mosaic, South Gallery, Hagia Sophia, Istanbul. © Giamberto/ Art Resource, N.Y.*

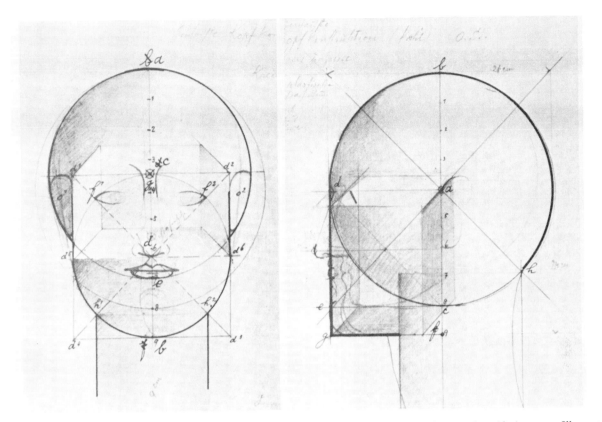

*Figure I.13* Two Studies of the Head Done by Oscar Schlemmer. *c. 1928. Courtesy Tut Schlemmer. [pg. 68,* Anatomy Illustrated *by Emily Blair Chewning. A Fireside Book, Simon and Schuster, New York, 1979.*

## A Study:

**Oskar Schlemmer comes very close to using the 3-circle scheme of Medieval times. The measurement from the root of the nose to its base is the same as from the nose to the chin. From the root of the nose to the crown is a little longer. He circumscribes the circle of the face within a square. In profile, Schlemmer simply moves his facial feature proportions across so that the viewer can see how the head is being developed.**

The *Gothic* (circa 12th to 16th century) figure no longer originates from the human being. The geometric schematic use of line traces the body attitude or stance of a figure, whereupon the figure is drawn within that scheme. The figure becomes decorative, not perfect physically or spiritually, but a supporting scheme for much church architecture.

Similarly, a student artist used calligraphic ratios from which to create figures.

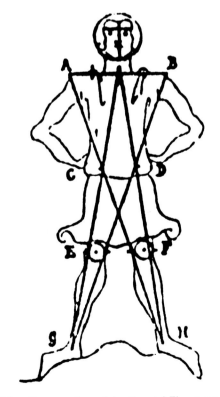

*Figure I.14* Construction of the Frontal Figure on the Basis of Villard de Honnecourt. *Paris, Bibliotheque Nationale, M. S. fr. 19093, fol. 19.*

129

## A Study:

Virginia Cofer asks us to note the small square units to the far left in the example below. Those units represent a pen nib width. The width can be narrowed or broadened with pen nibs of differing sizes, which then produce varying scales in figures. Here's how.

Each calligraphic alphabet has a prescribed number system for the ascender (the portion of the letter moving up), body (the main portion), and descender (the portion of the letter moving down), such as 3/5/3.

By applying the ascender/body/descender unit system to the human form you can end up with a figure that would vary according to the prescribed scale. The most interesting factor about this system is the infinite number of figures that are possible. While the scales are prescribed, the artist is left to make choices.

Notice the "Blackletter" ratio is 2/5/2, the "Uncial" 1.5/5/2, and the "Humanistic Minuscule" is 3/5/3. The artist's arbitrary systems are on the bottom line.

Alongside concerns about proportioning the human figure, artists looked for centuries for ways to make space appear *real* on a flat plane. Linear perspective is understood to be first identified by Brunelleschi, an architect in Florence, Italy, c. 1420. But another Italian named Alberti tried to set standards for a new convergence, the perfect human form *within* a perfect space. He devised something like a lengthening/retracting measuring stick calling it the *Exempeda*. The stick used a unit of six lengths. If an artist needed a smaller figure in space, the six units would be scaled down. If he needed a tall man, the units would be lengthened, all the while keeping width and diameters shifting to *ideal* proportions. Interestingly, Alberti had selected his proportional information from a number of models he considered ideal.[10]

The illustration below shows the application of Alberti's principles of perfection.

You might have noticed through these illustrations that there is little, if any, motion seen. All the figures are static. We spoke earlier about the Egyptians ignoring three crucial aspects in the drawing of figures: movement of parts, foreshortening, and the viewer's position. Most of those issues were addressed in Greek, then in Italian statuary. Idealized statuary figures were poised in motion, then placed in appropriate architectural niches at varying heights for bystanders to view.

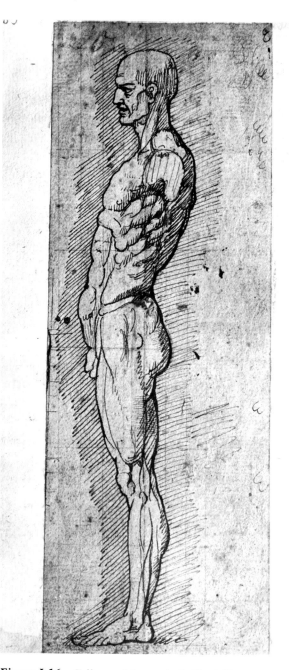

**Figure I.16** *Follower of Leonardo da Vinci. Figure proportioned according to L. B. Alberti's "Exempeda." Drawing in the codex Ballardi. Photograph. Giraudon, No. 260. Giraudon/Art Resource, N.Y.*

**Figure I.15** *Virgiania Cofer.* Using Calligraphic Number Systems to Create Figures.

Leonardo wrote on the subject of motion but his notes were not collated until *The Codex Huygens*, c. 1570, surfaced, ostensibly from Leonardo's statements on human movements. (No one knows who the author and draftsman of the book was, but the scholar M. W. M. Mensing states that not one single original drawing by Leonardo is in the manuscript.[11] Leonardo fused ideal human proportions with human movement and brought his theory into one principle, "the principle of continuous and uniform circular motion."[12] ". . . the fingers move by virtue of the hand, the hand by virtue of the arm, the arm by virtue of the body, and the body by virtue of the spirit."[13] Amazingly, his postulates on motion became the forerunners for the principles of motion pictures.

His conclusions are these:

1. "Every visible object can be seen from an infinite number of places . . ."[14]
2. Every bodily action is a succession of phases in space. Apart from identifying both optical and kinetic operatives, Leonardo also suggested a system to use in drawing the figure in motion. That system was based on circular motion, which infers the movement of a body in space on a flat plane.[15]

## A Study:

Russ Bitterman said this about his drawing: "The circle is used as a concept or principle for expressing movement, not in making a figure or movement itself. By combining a theory of movement with a theory of proportions, the figure has a transitional quality. The figure becomes more realistic. Together, these two theories are more abstract than older canons and yet they provide a more realistic figure as form."

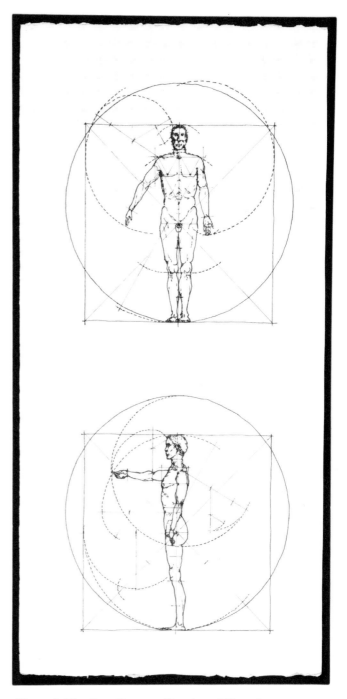

***Figure I.17*** *Codex Huygens, fol. 8 (13.5 by 18.4 cm.) Pierpont Morgan Library, N.Y. MA1139.*

***Figure I.18*** *Russ Bitterman.* Drawing of Vitruvian Proportions and Leonardo's Concept of Motion.

**Figure I.19** *Jules Etienne Marey (1830–1904).* Man Pole Vaulting. *A chronophotograph, c. 1890. Musee Marey, Beaune.*

The precursor to the Marey Wheel and motion picture was Leonardo.

But Jules Marey, the French physiologist, laid the fundamentals of cinematography in 1880 with photographic studies of motion. Unlike Muybridge's still frames of motion, Marey's pictures superimposed the stages of action in a single picture.

Another student abstracted the slow motion of a dancer. Movement involves time, the form the figure completes as it appears to move.

A contemporary of Leonardo's from the north, in Germany, was the last great master to turn his attention to systems for drawing the perfect figure. Albrecht Durer (1471–1528), from Nuremberg, was by far the most prolific chaser of the ideal body through numeration and proportion.

Durer's studies of proportions developed in five phases:

1. The earliest studies, geometrical constructions, 1500–1501.
2. The measurement of size and proportion (meaning anthropometric head lengths and fractions of the body length), c. 1505.
3. Progressive proportions, c. 1513
4. The "Exempeda Method," c. 1515
5. A stereometric method (meaning measurement of volumes) for use with figures in motion, c. 1523.[16]

Short-hand, rule-of-thumb applications still apply some of Durer's information in learning to draw, practicing with figures that are *eight heads high,* for instance. The middle of the body as the base of the pubic bone, another. Our

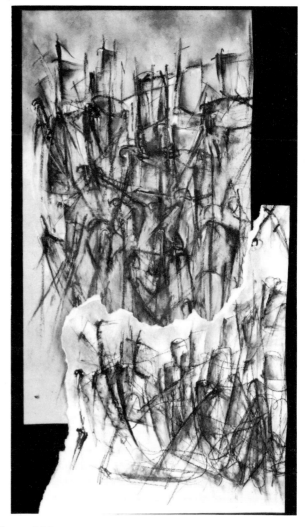

**Figure I.20** *Kris Lucas.* Abstraction of a Moving Figure.

*Colorplate 1*    Byzantine Christ *(c. thirteenth century). Detail of the "Deesis" Mosaic. South Gallery, Hagia Sophia, Istanbul. © Giamberto/ Art Resource, N.Y.*

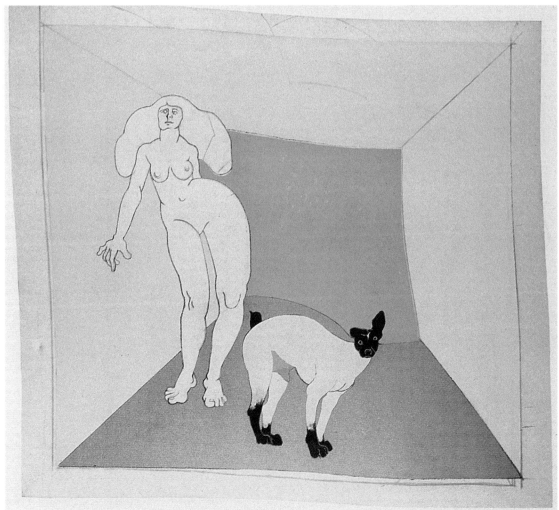

*Colorplate 2*    *Enrique Castro-Cid.* Flora and Benjamin. *1980. Acrylic on canvas, 61½″ × 65″. Collection Robert S. Cahn and Nancy Weber. Photograph by Scott Bowron. [Seen on page 53 of* Digital Visions, Computers and Art *by Cynthia Goodman. Harry N. Abrams, Inc., Publishers, N.Y.*

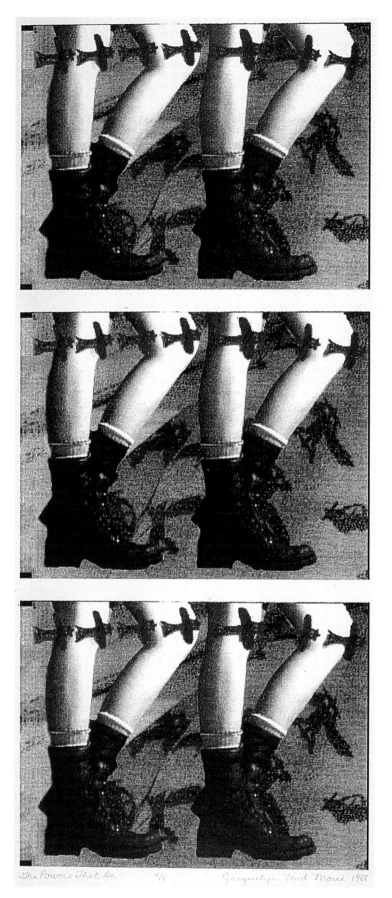

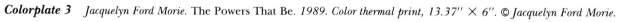

**Colorplate 3**  *Jacquelyn Ford Morie. The Powers That Be. 1989. Color thermal print, 13.37″ × 6″. © Jacquelyn Ford Morie.*

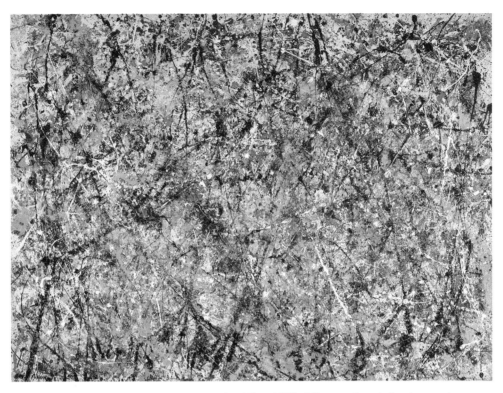

**Colorplate 4**  *Jackson Pollock (1912–1956). Number I Lavender Mist. 1950. Oil, enamel, and aluminum paint on canvas, 87 × 118 (2.210 × 2.997). National Gallery of Art, Washington, Gift of R. Horace Gallatin.*

**Colorplate 5**  *Willem de Kooning (1904– ). Woman I. 1950–52. Oil on canvas. 6'3⅞" × 58". Collection, The Museum of Modern Art, New York. Purchase.*

**Colorplate 6**  *Mel Ramos. I Still Get a Thrill When I See Bill. 1976. Oil on canvas, 80" × 70". Inventory #117. Courtesy Louis K. Meisel Gallery, New York. Photo by D. James Dee.*

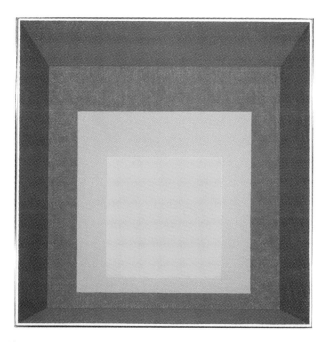

**Colorplate 7** *Josef Albers (1888–1976).* Homage to the Square: Apparition. *1959. Oil on board, 47½″ × 47½″. The Solomon R. Guggenheim Museum, New York. (Seen in Arnason's* History of Modern Art. *Colorplate 148.)*

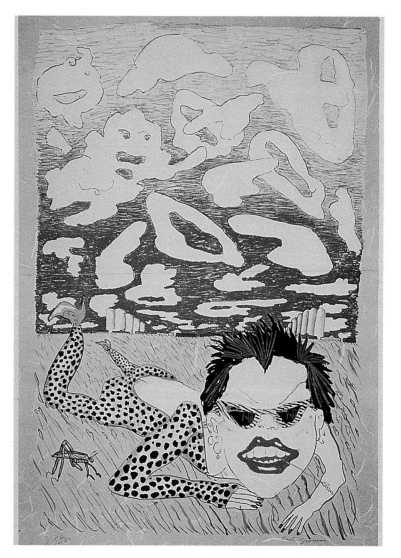

**Colorplate 8** *Red Grooms.* Lorna Doone. *1979–1980. Color lithograph with collage and rubber-stamp impressions printed on 2 sheets. 24½″ × 32″ each. Edition of 48. Collection: Brooke Alexander Gallery, Inc., New York, NY. Photographed by Eera-inkeri.*

**Colorplate 9** *Ed Paschke.* Lucy. *1973. Oil on canvas, 60¼" × 38" (153 × 96.5 cm) Museum of Contemporary Art, Chicago, Gift of Albert J. Bildner.*

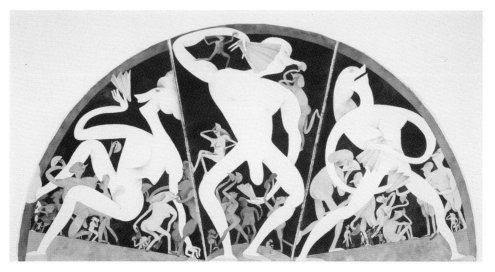

**Colorplate 10** *Gladys Nilsson.* Pandemoneeum—A Trip-Dick. *1983. Watercolour on paper. 51" × 101¼". (129.5 × 257 cm). Photo courtesy Phyllis Kind Galleries, Chicago and New York. Photo credit William H. Bengtson. Private collection.*

***Colorplate 11*** *Jean-Michel Basquiat. Untitled. 1981. Mixed media on wood panel, 73¼″ × 49¼″. Collection Robert Lehrman, Washington, D.C. (Seen in* Arts Magazine, *Vol. 64, No. 7, February, 1990. Illustration on page 55.)*

***Colorplate 12*** *Luis Cruz Azaceta. (Cuban, 1942– ). Homo Fragile. 1983. Acrylic on canvas, 72¼″ × 120¼″. (182.5 × 304 cm). Archer M. Huntington Art Gallery, The University of Texas at Austin, Archer M. Huntington Museum Fund, 1987. Photo credit: George Holmes.*

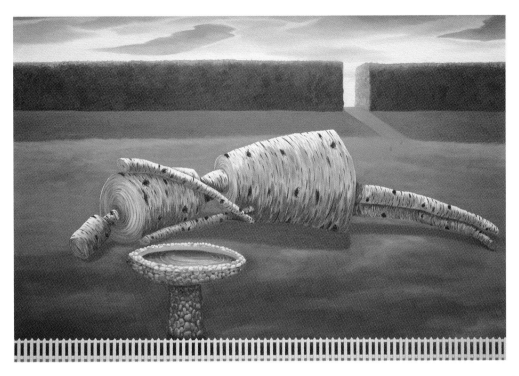

***Colorplate 13*** *Cheryl Laemmle.* Empty Birdbath. *1983. Oil on canvas, 56″ × 84″ (142 × 213 cm). Courtesy: Sharpe Gallery, New York. (Seen in* American Art Now *by Edward Lucie-Smith, Phaidon Press, Oxford, 1985. Illustration 105 on page 66.)*

***Colorplate 14*** *Charles Garabedian.* Ulysses. *1984. Acrylic on canvas, 90″ × 66″. The Eli Broad Family Foundation. Santa Monica, Calif.*

**Colorplate 15** *Gregory Gillespie.* Myself Painting a Self Portrait. *1980. Mixed media on panel, 58½″ × 68 3/4″. Photo courtesy Forum Gallery.*

**Colorplate 16** *Eric Fischl.* Haircut. *1985. Oil on linen, 104″ × 84″. The Eli Broad Family Foundation, Santa Monica, California. Photographed by William Nettles.*

**Colorplate 18** *Audrey Flack. Marilyn (Vanitas). 1977. Oil over acrylic on canvas, 96″ × 96″. Collection of the artist. Terry Allen. Photo, courtesy Louis K. Meisel Gallery, New York. Photo credit: Bruce C. Jones.*

**Colorplate 17** *William Beckman. Diana IV. 1981. Oil on wood panel, 84½″ × 50⅞″. Hirshhorn Museum and Sculpture Garden. Smithsonian Institution, Washington, D.C., The Thomas M. Evans, Jerome L. Green, Joseph H. Hirshhorn, & Sydney & Frances Lewis Purchase Fund.*

**Colorplate 19** *Dan Douke. Austin Healy Sprite. Acrylic painting. Nearly life size. Jack Glenn Gallery, Corona del Mar, California. (Seen in* Erotic Art of the Masters, the 18th, 19th, and 20th Centuries *by Bradley Smith. A Gemini-Smith, Inc. Book Published by Lyle Stuart, Inc., 120 Enterprise Avenue, Secaucus, New Jersey 07094.)*

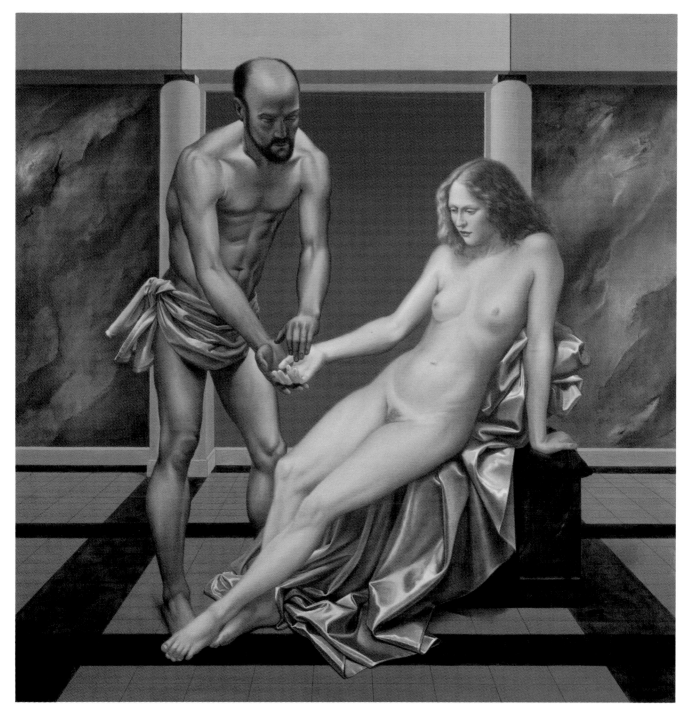

***Colorplate 20*** *Martha Mayer Erlebacher.* Mars and Venus.
*1983. Oil on canvas, 52″ × 52″ (132 × 132 cm). Courtesy of J.
Rosenthal Fine Arts, Ltd. Chicago, IL. (Seen in* American Art
Now, *by Edward Lucie-Smith. Phaidon Press, Oxford, 1985.
Illustration no. 184 on page 103.)*

*Colorplate 22* *Jennifer Bartlett. Rhapsody. (continued).*

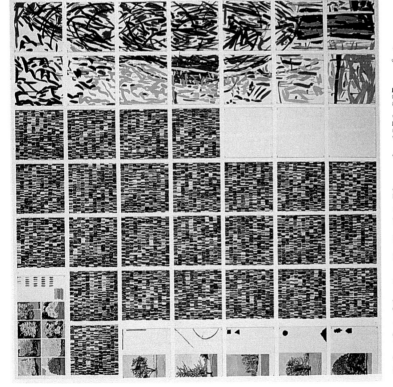

*Colorplate 21* *Jennifer Bartlett. Rhapsody. 1976. 987 one-foot-square steel plates baked with white enamel, silkscreened with a gray quarter-inch grid then painted on with enamel and a small brush, arranged in 142 rows with about 7 plates each. The work needs 153 running feet of wall for proper installation. (Seen in Rhapsody—Jennifer Bartlett. Introduction by Roberta Smith with notes by the artist and photographs by Geoffrey Clements. Harry N. Abrams, Inc., Publishers, New York, 1985. Illustrations excerpted from page 39, 40, and 41 (rows 85 through 105).*

**Colorplate 23**   *Graham Sutherland.* The Crucifixion. *1946. Oil on hardboard, 96″ × 90″. Church of St. Matthew, Northampton, England. Art Resource, N.Y.*

**Colorplate 24**   *Stanley Spencer, R.A.* The Crucifixion. *1958. Oil on canvas, 216 × 216 cm/85″ × 85″. Private collection.*

**Colorplate 26** *Robert Lentz.* Icon of Kateri Tekawitha. *1986. Gessoed untempered masonite, acrylic paint, 23 karat gold leaf. Bridge Building Icons.*

**Colorplate 25** The Virgin of Vladimir. *Byzantine. (ca. 1131 A.D.) 104 × 69 cm. Jahsrundert Tretyakov Gallery, Moscow, Russia. Art Resource, N.Y.*

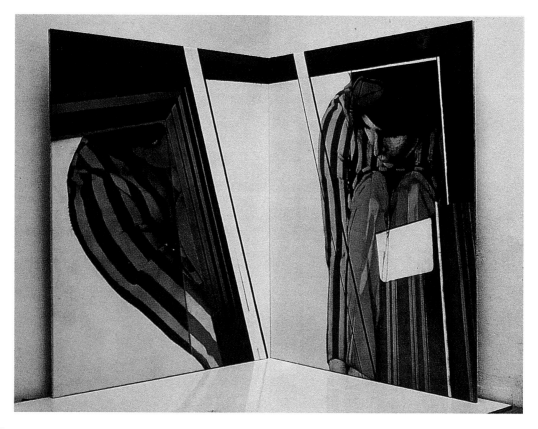

**Colorplate 27** *Pol Mara.* Out of the Corner of One's Eye. *1968. Oil on canvas and aluminum montage. 2 panels 195 × 162 cm each. Private collection, Los Angeles.*

**Colorplate 28** *Tom Wesselmann (American, 1931– ).* Great American Nude #35. *1962. Oil, polymer, and mixed media on board, 48″ × 60″ (121.9 × 152.3 cm). Signed on reverse, upper left: GAN #35/4′ × 5′ Wesselman '62. Virginia Museum of Fine Arts, Richmond Gift of Sydney and Frances Lewis, 85.454.*

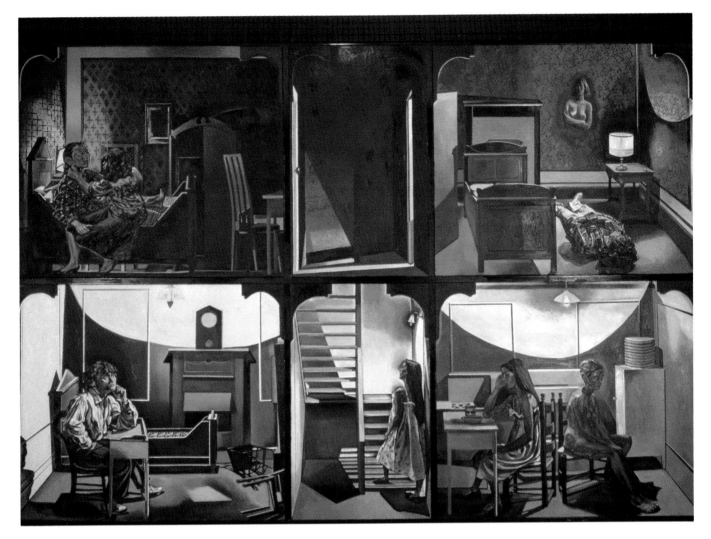

**Colorplate 29**   *Gabriel Laderman.* The House of Death and Life. *1984–85 O/C, 93″ × 135″. Schoelkopf Gallery, New York. (Seen in* Art in America *April, 1986, p. 172.)*

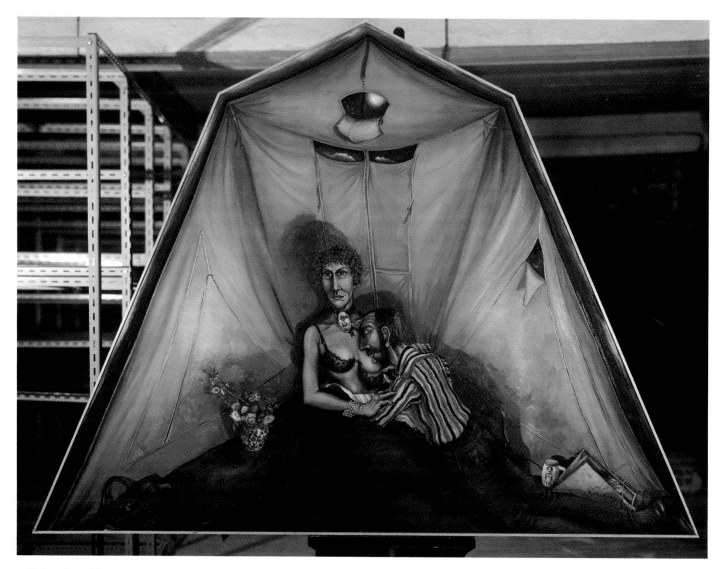

**Colorplate 30**    *Anthony Green.* Our Tent. Fourteenth Wedding Anniversary. *1976. Oil on board, 72" × 96". Collection: Rochdale Art Gallery. (Seen in the* Narrative Paintings Catalogue, *Arnolfini/Clive Adams Exhibition. The Arts Council of Great Britain.)*

hand length equals our face length, as well as the length of the scapula, clavicle (collarbone), and the sternum (breast bone). Other convenient measurements are these: two head lengths equals kneecap to outer ankle, elbow to fingertip, hip joint to kneecap. A child, age one, is four heads high, an adult is eight heads high, and in old age, seven heads high.

Our example from Durer's First Phase, geometric construction, has no fewer than 12 different circles or arcs defining measurements. (Finding the center of the circles laid out was an interesting enterprise. Often the centers rested on the points of intersection of two arcs.)

## A Study:

**Tim Greenzweig tried creating a figure by purely mechanical means just using a compass. Note the compass points in the example. Adjust the compass arm to create circular intersection points for proportions. Once you've found your proportions for height and width, complete the figure by drawing in the lines of the body as the student has done below.**

Another geometrical construction in Durer's First Phase shows how to construct a head.

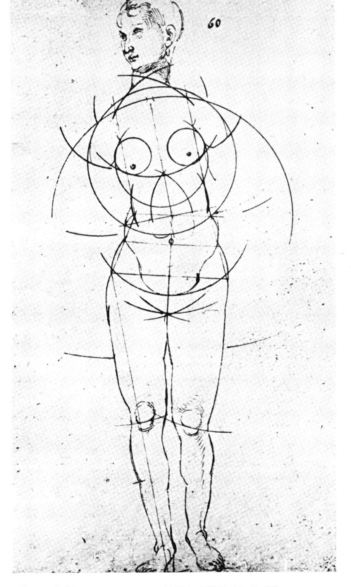

***Figure I.21*** *Albrecht Durer (1471–1528).* Nude Woman. *Constructed c. 1500. Page from the Dresden Sketchbook. Pen and ink with compass. 290 × 188 mm. (11 3/8″ × 7 3/8″) Dresden, Sachsische Landesbibliothek.*

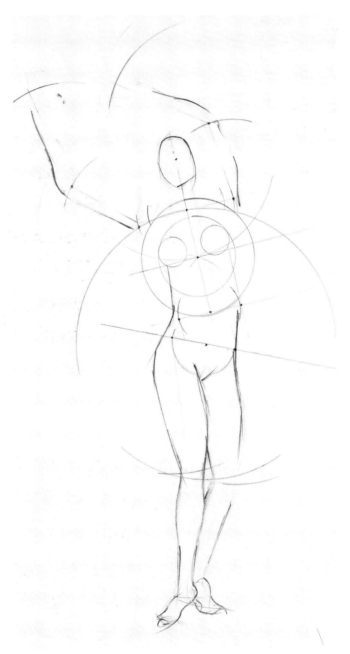

***Figure I.22*** *Tim Greenzweig.* Figure Created by Using a Compass.

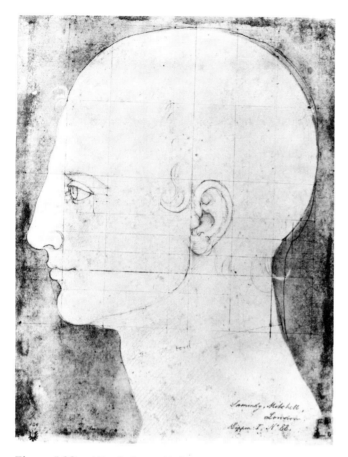

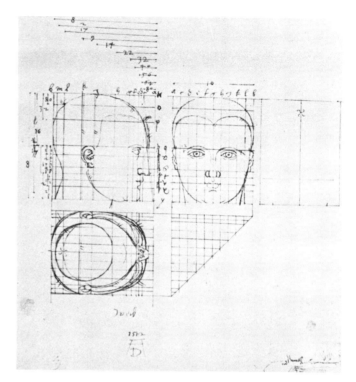

**Figure I.23** *Albrecht Durer (1471–1528).* Man's Head in Profile, Constructed Without Monogram. *c. 1504. Pen with brown and red ink, 243 × 289 mm. New York. J. Pierpont Morgan Library. New York. I, 257b verso. From collections of Mitchell, Lanna and Morgan.*

## A Study:

**Begin with a square. Divide the square with horizontal lines into thirds, thus identifying the hairline, the eyebrows, the base of the nose, the base of the chin. Then, extend the square on two sides by 1/6 of the original square to incorporate the nose and the back of the head.**

The Second Phase of Durer's preoccupation with human proportion tested the Vitruvian axiom of a man's body, 8 head lengths. But from this phase Durer added to measurements and proportions something he called "The Transfer Method."

## A Study:

The system Durer employed might be useful for you to reenact, perhaps using graph paper, though of course you can draw your own graph. The method is logical and very simple, and particularly helpful in foreshortening. You will gain some very good information about heads, faces, features, and foreshortening.

The head is still divided into thirds, hairline-eyebrow-nosebase-chin. Interestingly, the width of the head in frontal

**Figure I.24** *HP:1512/26 Ru. II/304; Bruck 116; SDS. 110. Head Constructed by the Transfer Method. No watermark. [pg. 2469 of* The Complete Drawings of Albrecht Durer, Volume 5, Human Proportions *by Walter L. Strauss, Abaris Books, N. Y., 1974].*

view, ear tip to ear tip, is equal to the face length, hairline to chin, making a square unit. (Leonardo had said that from the attachment of one ear to the other equaled eyebrows-to-chin[17]).

Once the graph on the right is constructed by extending the lines of information from the profile, moving left to right, draw the marks of brow, nose and chin, *frontal view.* The width of each of the features can be deduced by looking at the Durer drawings. Note that the width of the nose and mouth is the same as the distance between the inner corners of the eyes.

Now, the measurements of the foreshortened head are drawn from both the profile and the frontal views. Drop the units for ear, nose, mouth, etc. below from the *profile.* Drop the *frontal view measurements* into the triangular-shaped graph below. Those measurements in the triangular graph then need to move left for aligning the features.

You probably have noticed that because this is a constructed head and, unlike our own heads, absolutely symmetrical, the triangular grid below (right) can be used to *reflect* the placement of features from opposite sides. In other words, the line for the man's right ear descends to the lowest side of the foreshortened head below, which becomes his left ear, and so on for the placement of the eyes and such. The triangular graph lines stop at the points where, when taken left, meet the lines dropped from the profile to identify that special facial feature in foreshortening.

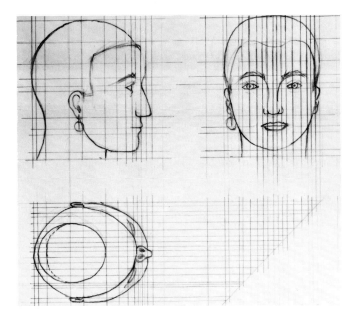

Figure I.25 *Tim Greenzweig.* The Transfer Method.

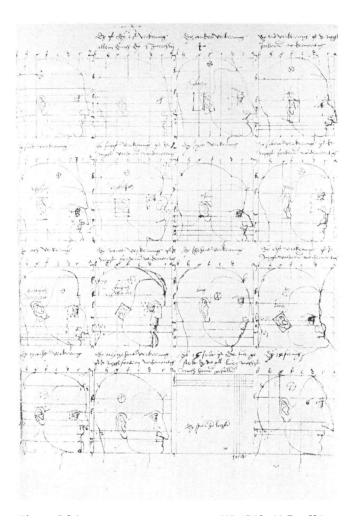

Figure I.26 Studies in Physiognomy. *HP:1513/41 Ru. II/ 472; Bruck 122; SDS. 117 [pg. 2498 in Strauss'* Durer, Volume 5].

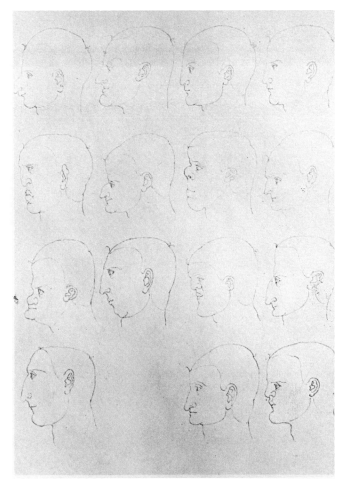

Figure I.27 Studies in Physiognomy. *HP:1513/42 Ru. II/ 472; Bruck 118; SDS. 118 [pg. 2498 in Strauss'* Durer, Volume 5].

Durer's Third Phase, c. 1513, included two types. Durer described several figures, among which was a child of four head lengths, assumed to be taken from Pomponius Gauricus' work *De Sculptura* (Florence, 1504).

Durer also used Gauricus' work as a springboard for his *Studies in Physiognomy*,[18] " . . . meaning the face, or countenance, especially when considered as an index to character."[19]

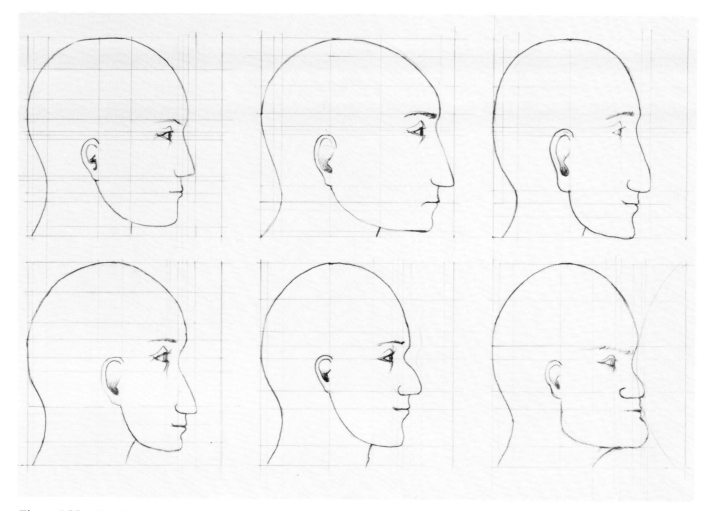

***Figure I.28*** *Tim Greenzweig.* Varying Ratios for Facial Features.

## A Study:

**You might want to try some of these transfers yourself, varying proportions for foreheads, eyes, nose, mouth, and chin. The study reminds you that a model's or a friend's features really can be captured in a drawing using simple proportional measurements. Many students are frustrated drawing faces. The transfer method gives you some rudimentary starting information, and is fun when features are varied.**

**Sean Miller drew two self portraits. One used elongated ratios.**

**The other used diminished ratios.**

The other type of work in this phase was *Progressive Proportion*, that is, parts of the body related to one another in width and length. What Durer was saying is that for the sake of strength, the hind parts of the body are longer and thicker than those in the front, the palm to the fingertips, the foot to the toes, and so on.[20] We will leave this phase of Durer's use of *The Transfer Method* with two illustrations of a hand and foot. You have already gone through the sequences of foreshortening, and seeing now what Durer does to turn the hand from top view to side view while retaining correct proportions should be relatively simple.

Durer says, "To draw the toes of a foot, do not have them extend straight out. But do as follows: the *big* toe should extend straight out, but the next three should be turned slightly inward, and one of them is to be a little longer than the big toe. The smallest toe should slant towards the center, as shown in the figure."[21]

Within his Fourth Phase, Durer created several devices to construct figures. One of these he called *The Divider* or *Variant*. Although Durer used harmonic proportions (see endnote 6) and set up both vertical (height) and horizontal (width) measurements, he still instructed the would-be artist that: "These parts or divisions thus indicated, you will divide according to your own pleasure . . ."[22] In so doing, Durer made very tall, thin figures and very short, fat figures.

Following Durer's methods with ruler, pencil, and graph paper can be interesting. Let me explain how some of the devices work.

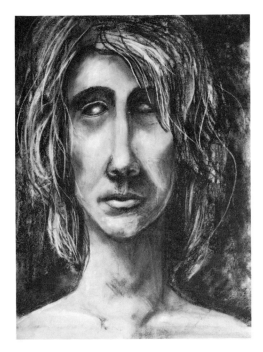

**Figure I.29**   *Sean Miller.* Self-Portrait with Elongated Proportions.

**Figure I.30**   *Sean Miller.* Self-Portrait with Diminished Proportions.

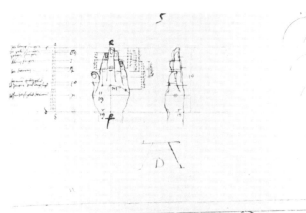

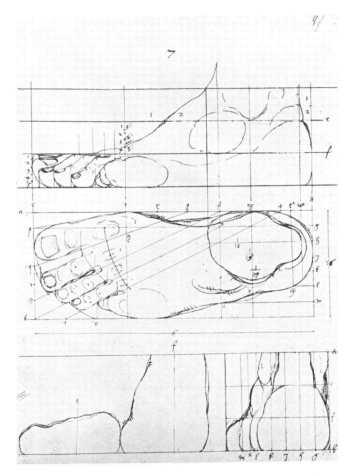

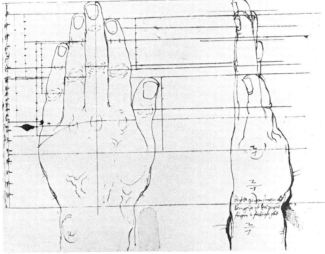

**Figure I.31**   *HP: 1513/48 Ru. II/320; Bruck 104; SDS. 123. "Right Hand and Left Hand." [pg 2501, Strauss'* Albrecht Durer, Volume 5].

**Figure I.32**   *HP:1513/56 Ru. II/330; Bruck 110; SDS. 125. "Left Foot in Profile," Top View, Rear View, and Cross Section [pg. 2505 in Strauss'* Albrecht Durer, Vol. 5].

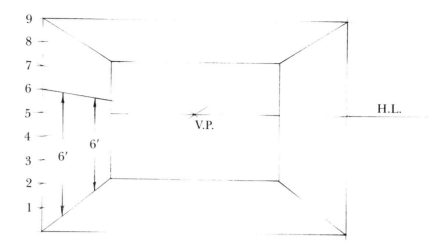

*Figure I.33*  *A One-Point Perspective Durer Called The Converter.*

## A Study:

–Draw a rectangle.

–Define nine equal units along the left side and number them, indicating feet from 0 to 9.

–Draw a horizontal line at 5 feet. That represents your Horizon Line (H L), or where the level of your eyes is.

–In the middle of the line put a dot. That is the Vanishing Point (V P).

–From #1 and #9, on left and right sides, draw lines to the V P.

–Draw two horizontal lines arbitrarily placed, for ceiling and floor.

–Draw two vertical lines for walls, connecting them to ceiling and floor lines.

–Erase the lines on what is now the back wall.

–Draw a line from #6 to the V P but only back far enough to intersect the back left corner.

That line represents *six feet all along that wall.* As you can see, as the line above the H L recedes to the V P, it moves downward. The floor line below the H L recedes upward to the V P. In this room a person would appear to become smaller in stature and narrower as he or she moved away from us. Durer experimented freely with this kind of perspective, calling his device *The Converter.*

## A Study:

Using the one-point perspective system, draw a man eight heads high and keep him on the front of the picture plane. Your drawing makes the figure *taller,* assuming the widths and vertical ratios are intact. Now, move back into and along the same lines that appear to recede, just as you would move back into the schematic room. You can draw the same man, keeping the same heights and widths. But, in transferring your receding man forward to the frontal plane, you will see you have created a *short, squat figure.*

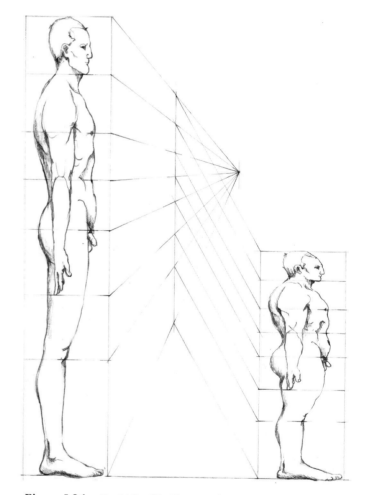

*Figure I.34*  *David Pavlik.* Figures of a Giant and a Dwarf. *Drawn by means of the "Converter."*

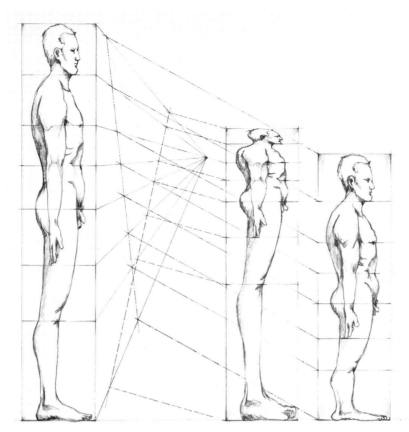

**Figure I.35** *David Pavlik.* The Falsifier, a Use of Angled Planes.

## A Study:

Another device Durer originated was *The Falsifier*. He constructed two types, and both used linear perspective. The first device prompted distortions by way of angled planes that shortened or elongated the tops or bottoms of the figure. Having a high or low horizon line made a very real difference, too. Again, it is worth an hour to enjoy this way of working. Just look at the illustrations, let the lines guide you, and see what you come up with.

## A Study:

Russ Nemec used a roving perspective for his architectural setting. He drew the model in class. Using his own scale, he then diminished the head and enlarged the hips, legs, and feet so that the reclining figure fits his ambiguous architecture.

**Figure I.36** *Russ Nemec.* Figure Scaled for a Roving Perspective.

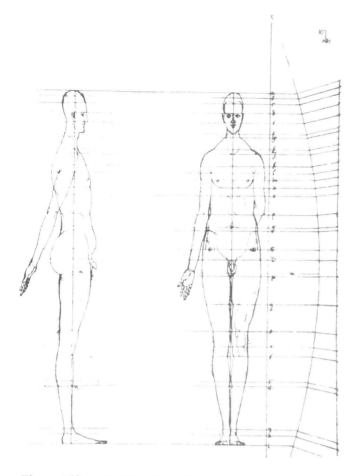

***Figure I.37*** *HP: 1526/23 Ru.II/496; Sloane 5228/107 r.*
Man of Nine Head Lengths, Altered by Means of the Second
''Falsifier'' *[pg. 2633 of Strauss' Durer, Vol. 5].*

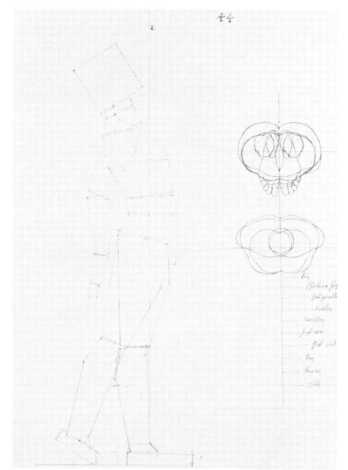

***Figure I.38*** *HP: 1527/18 Ru. III/157; Bruck 61; SDS. 101.*
Stereometric Man; Two Cross-Sections of the Body *[pg 2643 in
Strauss' Durer, Vol. 5].*

## A Study:

**A second *Falsifier* incorporated a convex or concave plane.**[23]
**Durer began with the original proportions in the
foreground, receding those measurement lines to an arc.
(The Horizon Line, in our example, now is at the halfway
mark of the figure, the beginning of the genitals.) He moved
lines left from the intersection points. He drew a figure
whose proportions now appear distorted.**

Also within this Fourth Phase, Durer harked back to
Alberti's *Exempeda* and set up a measuring stick of his own,
rather cumbersome and not very useful here.

His Fifth and final phase was a series of studies having
to do with movement and stereometry, which is the study
of volumes. Most of the drawings were done the year prior
to his death in 1528. The construction in our example used
a straight edge for lines. Note the extreme foreshortening

of two views, one from the feet and up, the other from the
neck and down.

With these drawings, the investigation of mathematics
and proportions of human beings came to a quiet end for
a long, long time. After two decades, Durer himself came
to believe that constructing an ideal figure might not be
feasible. But his systems for drawing figures probably
helped countless students learn to draw. Durer gave us very
useful tools—hints, insights, procedures, and constructs—to set us on our way. In our various shorthand
schemes, we still use them today.

Most of the men we have studied cautioned students
and followers that their *systems* were just guides that left
latitudes for choice. Corbusier, the twentieth century architect, may have said it best when he spoke of the neces-
sity of a *resulting diagram* (mathematics or geometry) which
*supplied equilibrium.* The artist then supplied the art.

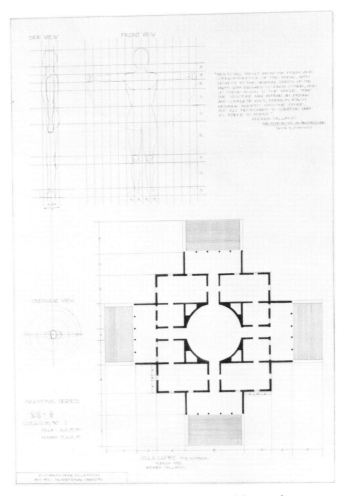

**Figure I.39** *Elizabeth Ellifson.* A Figure Drawn from Palladian Architectural Proportions.

## A Study:

**Andrea Palladio, an Italian architect, published four books on architecture in 1570.**

**Elizabeth Ellifson used the mathematical ratios of The Villa Capre to construct a figure. The quote she uses from Book I, Chapter I, sounds as descriptive for figure-as-form as for architecture.**

**"Beauty will result from the form and correspondence of the whole, with respect to the several parts, of the parts with regard to each other, and of these again to the whole; that the structure may appear an entire and complete body, wherein each member agrees with the other, and all necessary to compose what you intend to form."**

When systems based on mathematical proportions for the figure faded after Durer, the new academies such as Academia del Disegno, the first academy of fine arts, founded by Vasari, in Florence in 1562, taught geometry and anatomy alongside the regular studio workshops. The Academy of Painting and Sculpture, originated in France by Jean Baptiste Colbert in 1648, had twelve founding fathers, one of whom was Charles Le Brun (chief painter and perpetrator of the Louis XIV *style*). Le Brun published his own codes of human expression, with illustrations of people showing fear, awe, and so on. The codes were used for two centuries, especially by painters of mythological or historical subject matter.

In the academic orthodoxy that set in during and after the sixteenth century, the rational mind was challenged to reduce practice, appreciation, and taste to standard precepts. Not surprisingly, arguments arose over *draftsmanship or expression*, as if *both/and* did not count. In the seventeenth century, Poussin favored drawing, appealing to reason. Rubens preferred color, appealing to the senses. Ingre's nineteenth century teaching was an appeal for reason, whereas Delacroix, a contemporary, reiterated the battle cry for expression. And the cries are often no less faint in the art schools of our day.

**Figure I.40**  A Computer Mona Lisa.

**Figure I.41**  *Leonardo da Vinci, (1452–1519). Mona Lisa. c. 1503–05. Panel, 30 1/4″ × 21″. The Louvre, Paris. Scala/Art Resource, N.Y.*

But in the late twentieth century, some artists and technicians are drinking from the trough of mathematical technology once again. Frequently the technician grasps the mathematical complexities of the computer but lacks in history and practice the background of the fine artist. The result is often rich in graphic imagery but thin in content. Sometimes the artist who has years of expressive continuity behind him or her lacks the technical skill to bring electronic imagery to its highest potential. Ironically, what we are watching is mathematical construction once again trying to bring an *illusion of the real* into prominence. We could say we are *Back to the Future,* to use the name of the Robert Semeckis film of 1985. We are back with Brunelleschi and his vanishing point, back with Alberti and all the artists of the Italian Renaissance who tried to persuade us to see the *real* through illusion. The need to recreate nature is rising, using systems, this time with computers.

The original Mona Lisa is Figure As Form. She is an ideal, sufficient unto herself. She embodies a complete expression, an ordered concept. She embodies something understood to be beautiful and compelling. Her character represents *Woman,* not singular nor multiple but more than either, an archetype. We see in her what ought to be seen. Her original image instructs us completely.

A brief review of early computer experimentation with the figure will show that technical expertise exceeded scope and depth of figure-as-form, the search for the ideal. While the resulting images are idealizations, many are exaggerations, sometimes caricatures, simply more usable figurative representations than the Classical Greek ideal, which embodied the beautiful in nature and spirit. Indeed, understanding the computer's logic and language simply had to be the first point of order.

The computer can provide instant and extraordinary variations for building figures. An early concern for electronic imagery was fabricating a moving human figure. Bob Abel, in his construction of *Brilliance,* a robotic figure, began filming the motion of a dancer. On a computer workstation, the model's motions were recorded in such a way that the resulting animated figure appeared like a fluid 3-D figure.[24]

---

**142**

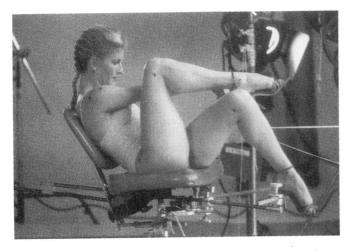

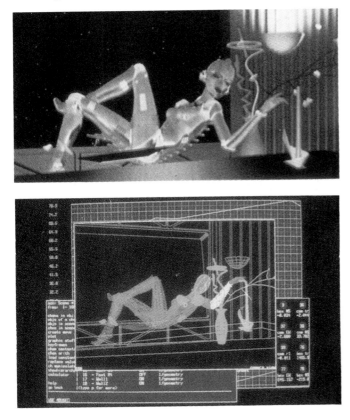

**Figure I.42** Dancer Filmed for Rendering the Motion of a Robot. *[From* Computer Graphics World, *July 1985, page 32. Pennwell Publishing Co., 119 Russell Street, Littleton, MA 01460. Abel Image Research Advertisement.*

**Figure I.44** *Frederic I. Parke.* Computer-Generated Head. *New York Institute of Technology Computer Graphics Lab [Old Westbury, New York 11568]*

**Figure I.43** Robot as Rendered on a Computer Workstation. *[From* Computer Graphics World, *July 1985, page 32. Pennwell Publishing Co., 119 Russell Street, Littleton, MA 01460. Abel Image Research Advertisement.*

Benoit Mandelbroth, at IBM's research center just outside New York City, created a new form of mathematics in 1975 called *fractal geometry.* Fractal geometry is a "category of shapes, both mathematical and natural, that have a fractional dimension. That is, instead of having a dimension of one, two, or three, theirs might be 1.5 or 2.25. A line, for instance, has one dimension and a plane has two, but a fractal curve in the plane has a dimension between one and two."[25] Fractals can approximate nature more closely in computer images with richer detail than other mathematical formulae.

Fred Parke, professor at New York Institute of Technology, pioneered facial animation in the 1970s. Sizes, shapes, and features were computer programmed in such a way that the numbers defining those units could be changed by the programmer.

Parke says that these images were created using a computer program that models the human face. Using "input parameters," he controls the shape and proportion of the face or regions of the face. And "if the parameter values (mathematical) input to the program are outside the normal expected value (mathematical) ranges, then distinctly unusual facial images may be created, such as the one shown in our example."[26] This means, for example, that if the eyes are moved further and further apart, eventually they extend beyond the limits of the face.

**Figure I.45** *Nadia Magnenat-Thalmann and Daniel Thalmann.* A Synthetic Actress, Marilyn Monroe. *University of Montreal, Canada. [Seen in* IEEE *Computer Graphics and Applications. December 1987. From the article "The Direction of Synthetic Actors in the Film Rendevouse a Montreal," by Thalmann and Thalmann. Illustration 1, page 9. Published by the Computer Society, 10622 Los Vaqueros Circle, Los Alamitos, California 90720] NAD/A Thalman.*

**Figure I.46** *Enrique Castro-Cid. Flora and Benjamin. 1980. Acrylic on canvas, 61 1/2" × 65". Collection Robert S. Cahn and Nancy Weber. Photograph by Scott Bowron. [Seen on page 53 of* Digital Visions, Computers and Art *by Cynthia Goodman. Harry N. Abrams, Inc., Publishers, New York]*

Among premier researchers in computer-generated figures are the Thalmanns, (Nadia Magnenat, wife, and husband Daniel). This couple worked for years in Montreal developing *synthetic actors*, Marilyn Monroe and Humphrey Bogart. The Thalmanns, now in Switzerland, program their figures not only as real-life correlatives, but enable them to speak and move as the original actors. This intensive computational task will ease with successively more sophisticated computer capabilities.[27]

Several artists who are experimenting with electronic imagery should be mentioned in passing, not because their work represents the *ideal*, because none of it does, but because their work shows the phenomenal current potential for expression through the use of numeration via computer technology.

Enrique Castro-Cid "spent ten years studying a branch of mathematics concerned with conformal transformations that preserve the angles between intersecting curves but not the shapes of the areas these curves enclose."[28]

Finding this too awkward, Castro-Cid and mathematician Robert Cahn turned to a mainframe computer. His images now begin with an original drawing. The drawing is digitized. "Once the X and Y coordinates are stored in the computer, the composition is transformed according to the mathematical equations of the program, then plotted. Any number of variations of an original drawing can be created in this manner simply by changing the equation. The final image is either developed as a drawing or projected onto canvas and painted."[29]

Jacquelyn Ford Morie, a computer artist and editor of *The American Journal of Computer Art in Education*, finds the figure an important subject in her work.

Usually she works with someone she knows personally. "With the computer, I can extract the essence of a figure and make it more universal. I can digitize the figure, bringing it into the computer by means of a video camera, which changes the video signal into individual pixels, or colored dots of light on the computer screen. These dots of light can then be manipulated, through the means of unique computer tools, and the figure can evolve into a picture with a life of its own."[30]

For *The Powers That Be*, Morie digitized her daughter's legs and saved the file in the computer. During another session she digitized her own red sweatshirt that had a black airplane pattern. Using a device called a Time Base Corrector, Morie boosted the video color to a supersaturated point. The airplanes turned orange, and the "noise" (heightened colors) in the background was intensified. This file was saved too.

A composite of the legs and sweatshirt was made. Isolating the small airplanes with less boosted color, Morie repeated the airplane shape across the top of the work, drawing shadows beneath the planes to finish the image.

*Figure I.48* *Kay-Lynne Johnson.* Horizontal Portrait. *1990. Computer and laser scanned image.*

## A Study:

**Kay-Lynne Johnson's** *Horizontal Portrait* **was produced using a laser scanner and computer. Through a network of graphed points, the artist touched her face with an electronic wand. Then, using her own additive process, the artist transferred her image to the laser scanner. Once in the scanner, the artist took control of the graphed points. Manipulating and merging images, she found she could create a variety of related images.**

This paradigm, *Figure As Form,* began with an exploration of the concept of an ideal figure. More often than not, the philosophers tracked that exploration to mathematical structures. But as we have seen, geometric figurative structures do not in themselves entirely achieve harmony and beauty. Exact symmetry often breeds boredom. The beautiful, embodied in the ideal, finally rests within esthetic interpretation by the artist. However, computer technology now allows us again to transform figure to numeration and numeration to figures. The new figures are based on the concept of the *ideal* only insofar as an ideal mathematical mode is called upon. The computer artist has a phenomenal tool to make images, figurative or not.

In *Figure Against Form* we will observe how some artists are not interested at all in the ideal beauty of a figure, but seek to express forms alone, in which a figure's presence, when seen, takes shape more from the artist's private feelings, rarely as an analogue to nature. Computer artist Steve Herrnstadt's work is a case in point. About his series titled *Phoenix,* he says, " . . . The Phoenix series is what could be called metaphysical self portraits. Rather than interpret and transcribe the outward appearance of the human form, the images are drawn from inside, the spiritual side. Spiritual here is not a religious term. The twelve

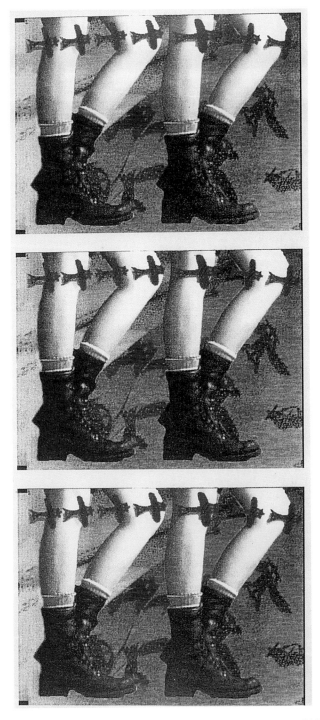

*Figure I.47* *Jacquelyn Ford Morie.* The Powers That Be. *1989. Color thermal print, 13.37″ × 6″. © Jacquelyn Ford Morie.*

"For the final print, I duplicated the screen image three times on a color thermal printer, which uses a wax-like dye melted onto a paper substrate. I used the repetition to depersonalize the figure, to make a more universal statement . . . referring to military powers (who originally enabled computer systems) and the power of teenage girls."[31]

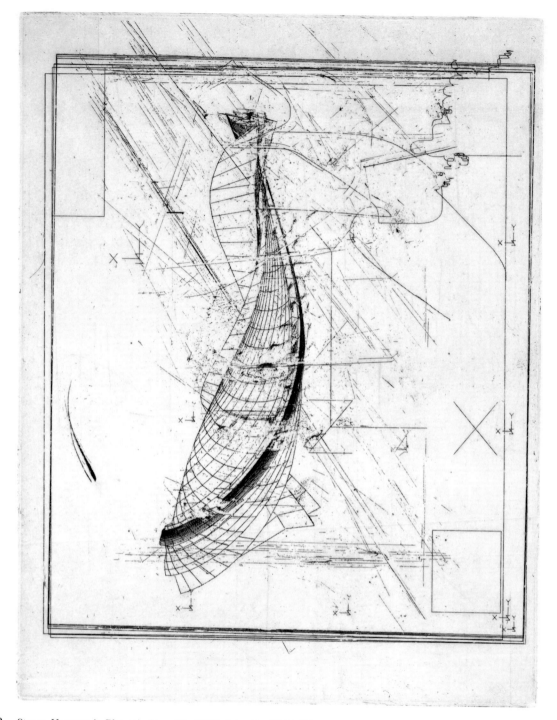

***Figure I.49*** *Steven Herrnstadt.* Phoenix Nov 03. *1988. Intaglio, 9″ × 7″.*

images are my journey about me, the figure, and how it feels to be who I am.

"The series was created on a VAX 785 computer using a finite element analysis software. The resulting images were transferred to an intaglio plate by machine for final printing by hand . . . I find my best work is usually a hybrid through the interaction between the more conventional media (drawing, painting, printmaking, photography) and the most recent technology I can contact."[32]

## ENDNOTES

1. Rudolf Wittkower, 1978. *Idea and Image Studies in the Italian Renaissance.* New York: Thames and Hudson, Ltd., p. 123.

2. Erwin Panofsky, 1955. *Meaning in the Visual Arts.* New York: Doubleday, and Company, Inc., Doubleday Anchor Books, 1955, p. 60.

3. Interview with Christene Thelen Fall, 1990.

4.  Panofsky, *Meaning in the Visual Arts*, p. 60.

5.  J. J. Pollitt, 1972. *Art and Experience in Classical Greece*. Cambridge University Press: London, pp. 106–108.

6.  Proportion has to do with mathematics and geometry. Footnote examples *a.* through *i.* should be easy to follow with the step-by-step instructions. Working your way through the steps slowly will help you understand a few of the numerical processes so many artists used in their attempt to find and apply order and harmony, and thereby render the figure as a form.

Mathematics is a special kind of language. Its grammar consists of symbols:

Numerals for numbers

Letters for unknown numbers

Equations for relationships between numbers

Pi for the ratio of the circumference to the diameter of a circle

Square Root for a number multiplied by itself, which gives you its square (the square root of 25 is 5.)

Proportion: the comparative relation between things or magnitudes; ratio; a reduction to scale.

*Geometric Proportion*: 1:2:4:8:16.

The first term is to the second as the second is to the third, etc. The fraction or length can be an arbitrary choice, but the *RATIO* must remain constant.

For instance: if "1" is the longest length and the *constant ratio* was 1/2, you would divide successive strings (lengths) by 1/2. The divisions would be 1:2:4:8:16 or $1 : \dfrac{1}{2} : \dfrac{1}{4} : \dfrac{1}{8} : \dfrac{1}{16}$ .

OR: You could use "1" as the longest length with a *constant ratio* of 2/3 and the divisions would be $1: \dfrac{2}{3} : \dfrac{4}{9} : \dfrac{8}{27} : \dfrac{16}{81}$ .

You could also begin with the smallest units and work up to the longest lengths, but the RATIO must remain constant.

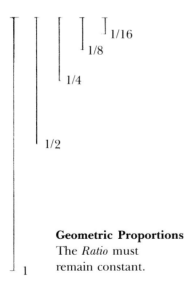

**Geometric Proportions**
The *Ratio* must remain constant.

*Geometric proportions.*

*Arithmetic Proportion*: 1:2:3:4:5. When the ensuing term exceeds the second by the same amount. The number, fraction, or ratio chosen can be arbitrary, but the DIFFERENCE *of actual lengths* must remain the same. You can start at either end, the longest or the shortest, but it is more natural in arithmetic proportion to begin with the shortest or smallest and use multiples of that fixed unit, *always maintaining a constant difference of actual lengths*. For instance: 5−4=1; 4−3=1; 3−2=1; 2−1=1;

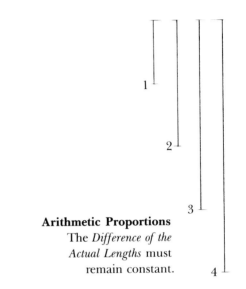

**Arithmetic Proportions**
The *Difference of the Actual Lengths* must remain constant.

*Arithmetic proportions.*

*Harmonic Proportion*: $1 : \dfrac{1}{2} : \dfrac{1}{3} : \dfrac{1}{4} : \dfrac{1}{5}$ .

A sequence of numbers whose inverses are in arithmetic progression. This means: the *inverses* of the above fractions are 1:2:3:4:5. The DIFFERENCE *of the inverses must remain constant*. You can start at either end, the longest or the shortest, but it is more natural to begin with the longest and divide the original string (lengths).

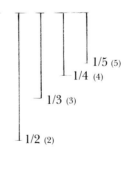

**Harmonic Proportions**
The *Difference of the Inverses* must remain constant.

*Harmonic proportions.*

For instance: $1 : \dfrac{1}{2} : \dfrac{1}{3} : \dfrac{1}{4} : \dfrac{1}{5}$ and $5 - 4 = 1$;
$4 - 3 = 1$; etc.

The following discussion of the Golden Ratio will be helpful in considering another of the constructs thought to be used in seeking and systematizing ideal proportions, from vases to persons to architecture. Because the "Golden Ratio" is so critical in the history of art, we will describe what it is and suggest how we may apply it to visual form.

Suppose we have a rectangle of sides "a" and "b" with "a" the shorter length. Now cut off a square of side "a" with all sides equal, leaving a smaller rectangle whose long side now is "a" and whose new shorter side is "c." We have $c = b - a$. If the RATIO of sides of the smaller rectangle is the same as for the larger, i.e., if $a/b = c/a$ or $a:b = c:a$, then this fixed ratio is the Golden Ratio. The magic number expressing this ratio (shortside, "a," divided by longside, "b") is .618.

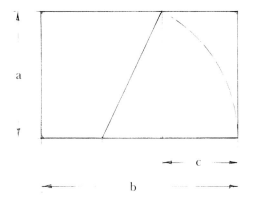

*A golden rectangle.*

The Greeks probably arrived at this ratio geometrically. We solve it algebraically with this equation $a = \dfrac{\sqrt{5} - 1}{2} = .61803$.
The square root of 5 ($\sqrt{5}$) is 2.236. Thus, $a = \dfrac{2.236 - 1}{2}$ which means $a = \dfrac{1.236}{2}$ which means $a = .618$, the numeric expression of the Golden Ratio.

Another way to put it is this:
$$\frac{\text{Short side}}{\text{Long side}} = \frac{\sqrt{5} - 1}{2} = \frac{2.236 - 1}{2} = \frac{1.236}{2} = .618$$

The application of the ratio to the Golden Rectangle is very simple:

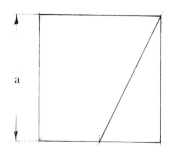

*A square with a diagonal to its half.*

Begin with a square and a diagonal to its half.

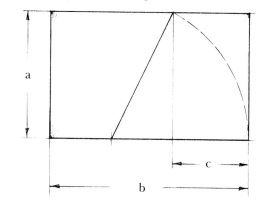

*Construction of a golden rectangle.*

Define the arc of a circle of the square to its half.

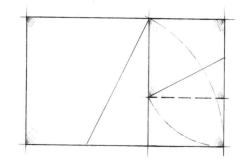

*A golden rectangle with another, smaller square inserted.*

Complete the rectangular unit on the right side of the square.
$$a:b = c:a \text{ OR } \frac{a}{b} = \frac{c}{a}$$

Within the rectangle you just constructed, you can again define a square whose relationship to the smaller rectangle will be the Golden Ratio again.
$$\frac{\text{Short Side}}{\text{Long Side}} = \frac{\sqrt{5} - 1}{2} = \frac{2.236 - 1}{2} = \frac{1.236}{2} = .618$$

If you continue this process within the Golden Rectangle, you will always retain that Golden Ratio between sides, .618.

The smaller is to the larger as the larger is to the whole. It is the only ratio that is also a proportion. Mathematically, in its simplicity, it is beautiful. And according to Adolph Zeising in 1854, the Golden Ratio was " . . . the perfect mean between absolute unity and absolute variety."

Leonardo of Pisa, called Fibonacci (1175–1230) discovered that if a ladder of whole numbers is constructed so that each number on the right is the sum of the pair on the preceding rung, the *ratio* between the two numbers approaches the Golden Ratio. That ratio can be approximated this way:

$1 : 2 : 3 : 5 : 8 : 13 : 21 : 34 : 55 : 89 : 144 : 233$, etc.

To arrive at the Golden Ratio, divide the large number into the smaller; for instance, $89 \div 144 = .618$. The Fibonacci sequence *tends toward* the Golden Ratio. If you built a stretcher for a canvas 34 inches by 55 inches, your rectangle would approach the Golden Rectangle.

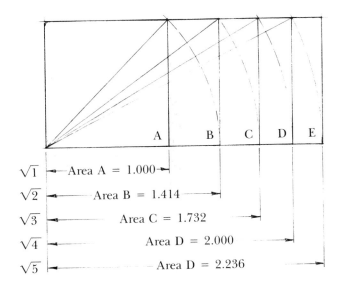

$\sqrt{1}$ |—Area A = 1.000—|

$\sqrt{2}$ |——Area B = 1.414——|

$\sqrt{3}$ |———Area C = 1.732———|

$\sqrt{4}$ |————Area D = 2.000————|

$\sqrt{5}$ |—————Area D = 2.236—————|

*The root 5 rectangle.*

We will turn now to look at an allied rectangle, the Root 5 Rectangle. The Root 5 Rectangle is easily constructed.

If you take the $\dfrac{\text{short side}}{\text{long side}} = \dfrac{1}{\sqrt{5}} = 0.477$ is the ratio.

In other words $\sqrt{5} = 2.236$ and $1 \div 2.236 = 0.477$.

There are arguments and counterarguments about the appropriation of the Golden Ratio, the Root-five Rectangle, and the Square to create "the beautiful."

What the argument for the use of the $\sqrt{5}$ rectangle has meant to the human figure is often, though seemingly erroneously, applied to figures like the Doryphorous in our example. We could produce a square if we took a measurement from his knees to the chest at the nipples. Then from the nipple to the crown, and from the knee to the foot would be .618 and .618. The square is 1.0, plus .618 plus .618 equals 2.236 or $\sqrt{5}$, a Root 5 Rectangle. Well, in the first place we have Roman copies of the originals. The extant writing of Greek proportioning is slim, and finally, no one really knows what "system" the Greeks used to achieve what they did.

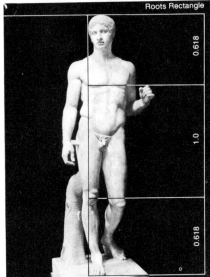

Roots Rectangle

0.618

1.0

0.618

*Doryphorous—The spearbearer. No one really knows.*

7. Pollio Vitruvius, *The Ten Books on Architecture*, trans. Morris Hicky Morgan, 1960. New York: Dover Publications, p. 73.

8. Leonardo da Vinci, *Notebooks of Leonardo da Vinci*, Vol. 1, compiled and edited by J. P. Richter, 1970. New York: Dover Publications, p. 182.

9. Panofsky, *Meaning in the Visual Arts*, p. 74.

10. Battista Alberti, Leon. *On Painting and Sculpture*, edited with translations, introductions, and notes by Cecil Grayson, 1972. London: Phaidon Press Limited, pp. 125–139.

11. Erwin Panofsky, 1940. *The Codex Huygens and Leonardo da Vinci's Art Theory*. Studies of the Warburg Institute, XIII, London, Westport: Greenwood Press, Inc., p. 10.

12. Panofsky, *Meaning in the Visual Arts*, p. 98.

13. Panofsky, *The Codex Huygens*, p. 24.

14. Ibid, p. 126.

15. Ibid, pp. 126–128.

16. W. L. Strauss, 1974. *The Complete Drawings of Albrecht Durer*, Vol. 5, *Human Proportions*, New York: Abaris Books, p. 2407.

17. Leonardo da Vinci, *Notebooks of Leonardo*, p. 171.

18. Strauss, *The Complete Drawings of Albrecht Durer*, Vol. 5, p. 2473.

19. *Random House Dictionary*, rev. ed. 1975.

20. Strauss, *The Complete Drawings of Albrecht Durer*, Vol. 5, p. 2475.

21. Strauss, *The Complete Drawings of Albrecht Durer*, Vol. 5, p. 2503.

22. A. Arnhem, & Chez Iean Ieanfz., *Les Quatre Livres D'Albert Durer* 1613, p. 1.

23. Strauss, *The Complete Drawings of Albrecht Durer*, Vol. 5, p. 2634.

24. Abigail Christopher, July 1985. "The Making of Brilliance," *Computer Graphics World*, p. 32.

25. Susan West, July/August 1984. "The New Realism," *Science*, p. 37.

26. Excerpted from a letter sent by Dr. Frederic I. Parke, Professor, Computer Graphics Laboratory, New York Institute of Technology, New York.

27. From correspondence by Jacquelyn Ford Morie, September 18, 1990.

28. Cynthia Goodman. 1987. *Digital Visions, Computers and Art*. New York: Harry N. Abrams, Inc., pp. 56–58.

29. Ibid.

30. Interview with Jacquelyn Ford Morie and from her written statement, September 18, 1990.

31. Ibid.

32. Interview with Steve Herrnstadt at Iowa State University, Ames, Iowa, August 8, 1990.

# BIBLIOGRAPHY

Alberti, Leon Battista. *On Painting and On Sculpture.* Edited with translations, introduction, and notes by Cecil Grayson. London: Phaidon Press Limited, 1972.

Brunes, Tons. *The Secrets of Ancient Geometry—and Its Use.* Vol. I. Translated by Charles M. Napier. Copenhagen, Rhodes, Pub. 1967.

Chewning, Emily Blair. *Anatomy Illustrated.* New York: Simon and Schuster, 1979.

Christopher, Abigail. "The Making of Brilliance." *Computer Graphics World,* (July 1985): 32.

Clark, Kenneth. *The Nude: A Study in Ideal Form.* New York: Doubleday, 1956.

*The Codex Huygens and Leonardo da Vinci's Art Theory.* London: Cambridge, Eng.: The Warburg Institute, 1940.

*Computer Graphics World.* Abel Image Research Advertisement, (July 1985): 49.

*Computer Graphics World.* (December 1985): 50.

Delacroix, Eugene. *The Journal of Eugene Delacroix.* Translated by Walter Pach. New York: Crown Publishers, 1948.

Dubery, Fred, and Willats, John. *Perspective and Other Drawing Systems.* New York: Van Nostrand Reinhold Company, 1972.

Durer, D'Albert. *Les Quatre Liures D'Albert Durer.* A. Arnhem, Chez Iean Ieanfz, 1613.

Durer, Albrecht. *The Writings of Albrecht Durer.* Translated and edited by William Martin Conway. New York: Philosophical Library, 1958.

Edgerton, Jr., Samuel Y. *The Renaissance Rediscovery of Linear Perspective.* New York: Basic Books, 1975.

Ghyka, Matila. *The Geometry of Art and Life.* New York: Dover Publications, Inc., 1977.

Goodman, Cynthia. *Digital Visions, Computers and Art.* New York: Harry N. Abrams, Inc., 1987.

Hambidge, Jay. *The Elements of Dynamic Symmetry.* New York: Dover Publications Inc., 1967.

*Idea, A Concept in Art Theory.* Translated by Joseph J. S. Peake. Columbia: University of South Carolina Press, 1968.

*The Ideas of Le Corbusier on Architecture and Urban Planning.* Edited by Jacques Guiton. Translated by Margaret Guiton. New York: George Braziller, Publisher, 1981.

Janson, H. W. *History of Art.* Englewood Cliffs, NJ: Prentice-Hall, Inc., 1962.

Kitzinger, Ernst. *Byzantine Art in the Making.* Cambridge: Harvard University Press, 1977.

Koschatzky, Walter, and Strobl, Alice. *Durer Drawings in the Albertina.* Greenwich: New York Graphic Society, Ltd., 1972.

Le Corbusier. *Modulor 2, 1955, Continuation of 'The Modulor,' 1948.* Translated by Peter de Francia and Ann Bostock. Cambridge: Harvard University Press, 1958.

*Leonardo, a Journal of the International Society for the Arts, Science and Technology.* Elmsford, New York: Pergamon Press, Inc., 1989.

Lyons, Lisa, and Friedman, Martin. *Close Portraits.* Walker Art Center, Minneapolis, 1980.

*A Modern Book of Esthetics, An Anthology.* Edited with introduction by Melvin Rader, 5th edition. New York: Holt, Rinehart and Winston, Inc., 1979.

*The Modulor, A Harmonious Measure to the Human Scale Universally Applicable to Architecture and Mechanics.* Translated by Peter de Francia and Anna Bostock. Cambridge: The M.I.T. Press, 1954.

*New Columbia Encyclopedia.* Edited by William H. Harris and Judith S. Levey. New York and London: Columbia University Press, 1975.

*The Notebooks of Leonardo da Vinci.* Vol. I. Compiled and edited by Jean Paul Richter. New York: Dover Publications, 1970.

Osborne, Harold, ed. *The Oxford Companion to Art.* Oxford: Clarendon Press, 1970.

Pach, Walter. *Ingres.* New York: Harper and Brothers Publishers, 1939.

Panofsky, Erwin. *Meaning in the Visual Arts.* Woodstock, New York: The Overlook Press, 1974.

Panofsky, Erwin. *Albrecht Durer.* Vol. One and Vol. Two. Princeton, New Jersey: Princeton University Press, 1980.

Panofsky, Erwin. *Renaissance and Renascences in Western Art.* New York: Harper and Row, 1972.

Panofsky, Erwin. *Studies in Iconology.* New York: Harper and Row, Harper Textbooks, 1962.

Pollitt, J. J. *Art and Experience in Classical Greece.* Cambridge, Eng.: Cambridge University Press, 1972.

Pope-Hennessy, John. *Raphael.* New York: New York University Press, 1970.

*Random House Dictionary.* Revised edition. Jess Stein, Editor. New York: Random House, 1975.

Roriczer, Mathes, and Schmuttermayer, Hanns. *Gothic Design Techniques, the Fifteenth-Century Design Booklets of Mathes Roriczer and Hanns Schmuttermayer.* Edited and translated by Lon R. Shelby. Carbondale: Southern Illinois University Press, 1977.

Strauss, Walter L. *The Complete Drawings of Albrecht Durer. Volume 2: 1500–1509.* New York: Abaris Books, 1974.

Strauss, Walter L. *The Complete Drawings of Albrecht Durer. Volume 5: Human Proportions.* New York: Abaris Books, 1974.

Swaan, Wim. *The Late Middle Ages.* London: Paul Elek, Ltd., 1977.

Thalmann, Nadia Magnenat and Daniel. "A Computer Created Marilyn Monroe." *IEEE Computer Graphics and Applications.* The Computer Society Publishers, Los Alamitos, California, 1987.

Vitruvius. *The Ten Books on Architecture.* Translated by Morris Hickey Morgan. New York: Dover Publications, Inc., 1960.

von Moos, Stanislaus. *Le Corbusier, Elements of a Synthesis.* Cambridge, Mass.: The M.I.T. Press, 1979.

Weitzmann, Kurt; Alibegasvili, Gaiane; Volskaja, Aneli; Chatzidakis, Mandis; Babic, Gordana; Alpatov, Mihail; and Voinescv, Teodora. *The Icon.* New York: Alfred A. Knopf, 1982.

Weitzmann, Kurt; Chatzidakis, Manolis and Radojcic, Svetozar. *Icons.* Conceived and coordinated by Ivan Ninic. The Alpine Fine Arts Collection, Ltd, New York, 1980.

West, Susan. "The New Realism." *Science,* (July/August, 1984).

Weyl, Hermann. *Space, Time and Matter.* Translated by Henry L. Brose. New York: Dover Publications, Inc., 1952.

Wittkower, Rudolf. *Idea and Image, Studies in the Italian Renaissance.* New York: Thames and Hudson, 1978.

Wright, Lawrence. *Perspective in Perspective.* London: Routledge and Kegan Paul, Pub., 1983.

# Paradigm II
# Figure Against Form

▼

**M**uch of the information in the first paradigm, *Figure As Form,* surrounded geometric constructs of the figure, partly because a divisible physical body can be assembled with continuity, drawn and rearranged. The nude as a constructed model becomes a repeatable form, generalized, like a manikin. But even though the Greeks used mathematics to try to find a measurable, mystical proportion for the figure, the *ideal* figurative form is not so simply graphed.

You remember that the art historian Rudolph Wittkower reminded us that there are seventeen species of symmetry. Bilateral symmetry is only one. Just when we begin to complain that our mouth is crooked, one eye is lower than the other, or our nose is too long, we are wishing for ideal physical features. We want something more pleasing, more symmetrical—and we long for a symmetry that can be reproduced geometrically.

The human being as an *ideal form* is much harder to understand and to draw than is the purely physical form. The ideal human figure does not imply imitation of one person but perfection, something that rests finally within the artist's imagination. The ideal human form implies wholeness in body, mind, and spirit. A good working definition of the ideal human form could be this: That which unites artistically in a single form all the excellencies found in nature in different individual forms of the same type or belonging to the same category. *The Ideal* thus aims to be more perfect than anything which can actually be observed but necessarily proceeds from the artist's own idea of perfection.''[1] The ideal human form cannot be captured just with numerical ratios, as we surmised in the prior chapter.

Nonetheless, nearly all postmedieval artists tried to find divisible ratios to show a figurative form of divine proportion, something resonating in spirit and beauty. Figure as

Form. That concept of the figure played on, in varying degrees, for five hundred years. The twentieth century has given us something else.

Now, with *Figure Against Form* we find figure pushing away from form and form pushing away from figure. The title works both ways—Figure Against Form is Form Against Figure.

When form is at its purest there is no reason to deal with the perception of some real analogue, like a figure. Pure form is abstract. Its relationship to subject is gone. Form has no simple relationship to the world *out there.* Form filters out the figure and moves to psychological venues of lines, colors, values, textures, shapes, and space (the *form*-al elements). Pure form moves into abstraction, where feeling is not identified with an object, an "objective correlative," like a figure.

Art not from visible objects but from sources spiritual and intellectual can find a simplified comparison in music. Music is created from the mind by ordering notes, symbols of sound. The composer knows he or she is done when the works *feels* complete, resolved. The same process is used in abstract art, though the means are different. Abstract art stretches from the lyrical to the geometric, and the arrangement of formal elements is what defines content.

Art *abstracted* from nature until the "correlative other" is lost is one thing. Art *generated solely by the artist,* "artifact art," abstract art is another. Let's look at some differences.

Theo van Doesburg, a Dutch artist and architect, founded a periodical called *De Stijl* in 1917. The contributing artists emphasized a need for abstraction and simplification.

Van Doesburg's own artistic effort bred abstractions by reducing appearances to generalizations. The "correlative other," the figure, is lost between the third and fourth

**Figure II.1** *Theo van Doesburg.* Compositie in Dissonanten. *1919. Oil on canvas, 63.5 × 58.5 cm. Oeffentliche Kunstasammlung, Basel, Switzerland. (Seen in* De Stijl: The Formative Years–1917–1922. *MIT Press. Cambridge, Massachusetts. 1982. Illustration #20.)*

image in Doesburg's sequence *Compositie in Dissonanten.* In this example there is a digression from the dramatic diagonals of the figure in the third piece, to the vertical and horizontal passages in the fourth piece, to the closed squares and rectangles of the last work.

While the last illustration shows a move from the natural image to complete abstraction, van Doesburg's drawing of a cow never quite leaves the "correlative other." We can still pick out lines informing our eye that a cow existed. We can do this simply because we have a reducible sequence from a cow before our eyes. If we saw just the eighth drawing, we would have some difficulty discerning an original or natural shape.

In the 1970s, Roy Lichtenstein worked a series reducing a bull. His image loses the correlative other about the fifth image, most certainly in the sixth.

You've been reading and working with the figure as form throughout the introductory exercises and in the last

paradigm, basically dealing with a constructed figure, albeit an abstract construction; however the end result was *a figure*, identifiable, understood to be a figure.

But throughout history, repetitions and cycles in art show movements to and from the figure. *To* the figure describes or represents the human being. *Away* from the figure celebrates the mind and spirit of the producer of the art. Arsen Pohribny comments: " . . . in each painting, indeed in the whole of the history of art, the powers of attraction of the two poles (representational and non-representational, or differently put: of imitation and abstraction) attract and repel each other."[2]

Before we discuss "artifact art" (abstract) art, let us turn briefly to our legacy of artists who abstracted from nature, meaning artists who began to play and experiment with light, with structure, with lines, with whatever formal concerns they had, more than they copied natural appearances.

**Figure II.2** *Theo van Doesburg. The Cow. Eight pencil drawings; numbers 1, 2, 4, 5, 6, & 7, each 4⅝″ × 6¼″; numbers 8 & 9 each 6¼″ × 4⅝″. Collection, The Museum of Modern Art. New York. Purchase Fund.*

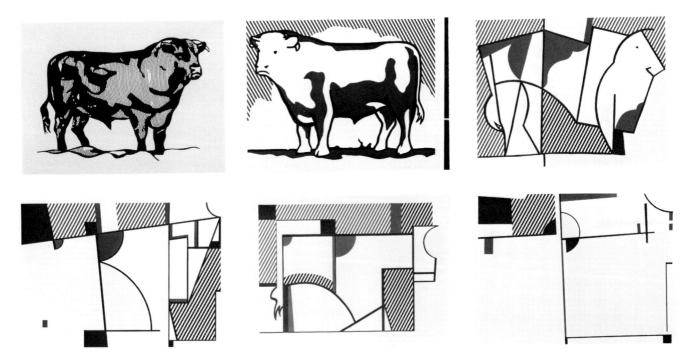

**Figure II.3** *Roy Lichtenstein. Bull Series. 1973. Private Collection. New York. "Bull I." India ink over dry marker. 28″ × 38.1″ (71.1 × 96.8 cm.). "Bull II." Collage and India ink. 27 1/4″ × 35.1″ (69.2 × 89.5 cm.). "Bull III." India ink and synthetic polymer paint. 28″ × 37 1/4″ (71.1 × 94.6 cm.). "Bull IV." Collage, India ink and silkscreen. 23 3/8″ × 33 1/8″ (59.4 × 84.1 cm.). "Bull V." Collage and India ink. 23 1/2″ × 33 5/8″ (59.7 × 84.1 cm.). "Bull VI." 1973. Collage, India ink, synthetic polymer paint, and pencil. 25″ × 33 1/4″ (63.5 × 84.5 cm.). [Seen in* Drawing Now *by Bernice Rose. The Museum of Modern Art, New York, New York, 1976. Illustrations on pages 44, 45, 46, and 47.] © Gemini G. E. L., Los Angeles, CA, 1973.*

**Figure II.5** *Paul Cezanne (1839–1906).* The Large Bather. *1898–1905. 82″ × 98″. Philadelphia Museum of Art. Wilstach Collection.*

**Figure II.4** *Edouard Manet (1832–1883).* The Fifer. *1866. Canvas. 63″ × 38 1/4″. Giraudon/Art Resource, N.Y.*

By the 1860s, Edouard Manet, a Frenchman, was asserting the picture plane as a two-dimensional surface, flattening figures by diminishing values and shadows, relinquishing chiaroscuro. The Renaissance picture plane was still there, but a painterly revolution was coming.

Paul Cézanne, another Frenchman, was less impressed with the vagaries of light and more impressed with the structure of nature. Shadows became shapes in and of themselves. Brush strokes were building blocks of warm and cool colors. He believed that all forms in nature were "... based on the cone, the sphere, and the cylinder."[3] Near the end of his life, Cézanne painted a large canvas, nearly 7′ × 8′, titled *The Large Bathers*. The figures were as formally structured as the vaulted trees. They became part of the architecture of the composition. Figure as form

resided here in the diminution of human individuality by moving the figure into strategic designs that appealed more to the mind than to the senses, thereby assuming a timeless, allegorical quality.

Braque and Picasso began to refract in still another way. Knowing that the eye focused only one place at a time, they reasoned that an object drawn from multi-views fractured the picture plane but would remain closer to the truth of the object drawn. Wasn't a figure drawn from all points of view a more informed drawing than drawing from only one point of view? Natural appearances became an abstract order that rested in a vertical, sculptural, bas-relief-like facade. With a narrow range of colors, an exclusion of representational light, atmosphere, and perspective, the Cubists sought to draw things as they *were*, not as they *appeared*. The multi-view fractured the object like a field of broken mirrors and the Renaissance window, illusion of the real, had exploded.

Three steps *of Cubism* were part of the radical legacy that transformed space for artists. From the early bas-relief-like facade of analytical Cubism that rested immediately *behind* the picture plane, to the use of words *on* the picture plane, to the collage that comes *in front of* the picture plane, spatial thinking was changing radically.

Picasso's *Still Life with Chair Caning* incorporated found materials. The materials were pasted onto the top of a shape, no longer the rectangular picture plane. The scraps did two things. The scraps *represented* part of an image within a composition, and they *presented* their own identity.[4] That bonding was a new perceptual construct.

These artists who abstracted from nature helped lay the groundwork for what followed.

The following diagram is a thumb-nail sketch of a very complex art history. But the *Form-Against-Figure-Against-Form Arc to the Nineties* illustrates some of the movements away from the figure, beginning in the 1850s, to the full bloom of Abstract Expressionism in the 1950s, to the current issues moving back to the figure in the 1980s and 1990s.

The twentieth century realms of *pure form* were shaped by several major artists. We will begin with two unknown

**Figure II.6** *Pablo Picasso (1881–1973).* Ambroise Vollard. *1909–1910. 36″ × 25 1/2″. Pushkin Museum, Moscow.*

**Form Against Figure Against Form**
**Arc to the Nineties**

Full Bloom of Abstraction
1950's - 1960's
Pollock - *Abstract Expressionism*
DeKooning
Albers - *Constructivism*

1930's　　Some "Mainline Outsiders"　　2000
Alice Neel, Alfonso Ossorio, Leon Golub, Red Grooms

"Transformers"
1960's - 1990's
Ed Pashke
Jim Nutt
Gladys Nilsson
Jean-Michel Basquiat
Luis Cruz Azaceta
Cheryl Laemmle
Jonathan Borofsky
Charles Garabedian
Gregory Gillespie
Eric Fischl
William Beckman
Audrey Flack
Martha Mayer Erlebacher
Jennifer Bartlett

1910's, 20's, 30's, 40's
Kupka - *Abstract*
Kandinski - *The Spiritual*
Malevich - *Suprematism*
Mondrian - *Plastic Art*
Van Doesburg - *De Stijl*

1900's
Braque & Picasso - *Cubism,*
*Simultaneous Instantenaity*

1890's
Cezanne - *The Cube, Cone and Cylinder*

1850's
Manet - *Flattening the Picture Plane*

**Figure II.8** Form-Against-Figure-Against-Form. *Arc to the Nineties.*

**Figure II.7** *Pablo Picasso (1881–1973).* Still Life with Chair Caning. *1911–1912. 10 1/2″ × 13 3/4″. Collection, the artist's estate.*

**Figure II.9** *Frantisek Kupka.* Disks of Newton (Study for Fugue in Two Colors). *1912. 39 1/8″ × 29″. Philadelphia Museum of Art. Louise and Walter Arensberg Collection.*

**Figure II.10** *Wassily Kandinsky (1866–1944).* Light Picture. *Dec 1913. Oil on canvas. 30⅝″ × 39½″. Solomon R. Guggenheim Museum, N.Y. Gift, Solomon Guggenheim, 1937. Photo: Robert E. Mates.*

**Figure II.11** *Kasimir Malevich.* Basic Suprematist Element. *1913. Drawing. Russian State Museums, Leningrad. [Seen in H. H. Arnason's* History of Modern Art. *Third Edition. Page 187.]*

to each other: a Czech, Frantizek Kupka, who worked in Paris, and a Russian, Vasily Kandinsky, who worked in Germany.

Factual motifs, that is, the real world, were a hindrance for Kupka. He was occupied with two basic elements, movement and color. "I don't think it is necessary to paint trees, as people can see better ones in the original on their way to the exhibition. I do paint, but only ideas, the synthesis; if you want, the chords."[5]

Among the earliest visual documents of abstract art rest in Kupka's *The First Step* and his *Disks of Newton (Study for Fugue in Two Colors)* 1912, a work with vibrating, pivoting circular color shapes, a geometrical element.

Kandinsky settled in Munich in 1890 and published his famous treatise, *Concerning the Spiritual in Art* in 1912. "The harmony of color and form must be based solely upon the principle of the proper contact with the human soul."[6]

Kasimir Malevich, another abstract pioneer, who studied art in Moscow said ". . . in the year 1913, in my desperate attempt to free art from the burden of the object, I took refuge in the square form and exhibited a picture which consisted of nothing more than a black square on a white field."[7]

This picture was a paradigm for the Suprematist movement. Malevich wrote, " . . . to the suprematist the visual phenomena of the objective world are, in themselves, meaningless; the significant thing is feeling, as such, quite apart from the environment in which it is called forth."[8]

Piet Mondrian left Holland for Paris about 1911. He progressed from naturalism through several "isms" including Cubism, to abstract art. He felt in 1942 that Cubism

" . . . did not accept the logical consequences of its own discoveries: it was not developing abstraction toward its ultimate goal, the expression of pure reality. I felt this could only be established by *pure plastics (plasticism).*"[9]

The *new reality* was in the plasticity of space (intense, bright colors appear to come toward one, while deep, dark colors appear to recede—yielding a surface elasticity) in the forms and colors on a canvas. The new reality was found in the presence of the painting itself.

Mondrian balanced unequal opposites, using the right angle for squares and rectangles. He simplified his colors to red, yellow, blue, black, and white.

Imitation and abstraction attract and repel each other. Figure is against form and form is against figure. As long as the figure is subject, its imitation, in whatever manner, will dictate to the artist. Creating the form as the subject itself, in whatever manner, will let the artist dictate to the work an expression of the human mind.

A word here for clarification and definition:

1. *Abstracting*—(a) the search for pure forms in nature. (b) the search for pure forms in the mind and spirit.
2. *Abstract Art*—understood to be a label for many types and kinds of nonrepresentational work. "Abstract Art is 'every kind of art which does not relate to visible reality' . . . it is art separated from all concrete objects . . . the aim is a concept of supra-individuality, 'objective art' . . . it is concrete exposition, precise, impersonal and immaterial; it is mostly based on geometrical forms which have a tangible concrete existence: 'for nothing is more concrete, more real than a line, a colour, a surface,' a 'cool' way of painting . . . trends of 'hot' abstractionism are typically represented by lyrical and abstract paintings."[10]
3. *Expressionism*—can be traced to the 1880s, to artists breaking with European Renaissance representation. Emotion and feeling were the only true goals of art through use of the formal elements to convey a more forcible statement, using a variety of means including distortion of subject matter.[11]
4. *Abstract Expressionism*—is considered to be a mid-twentieth century movement removed from geometric traditions. "The essence of Abstract Expressionism is the spontaneous assertion of the individual."[12] Two tendencies are suggested—*Action painters* are concerned with gesture and texture and *Color Field painters* are concerned with large, unified, abstract color shapes.[13]

Jackson Pollock began his huge action paintings about 1943 and ushered in Abstract Expressionism. One of Pollock's clarifiers, Harold Rosenberg, said: "The painting as

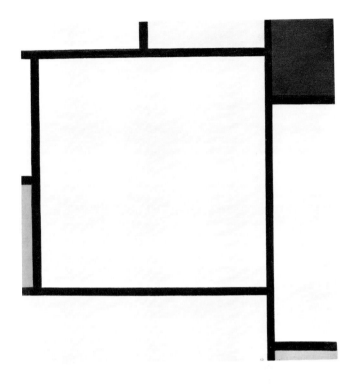

***Figure II.12A*** *Piet Mondrian (1872–1944).* Composition 2. *Solomon R. Guggenheim Museum, N.Y.*

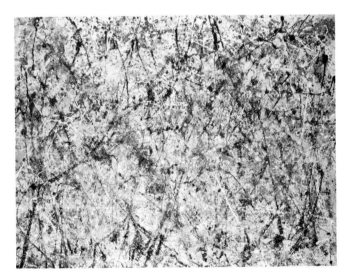

***Figure II.12B*** *Jackson Pollock (1912–1956).* Number I Lavender Mist. *1950. Oil, enamel, and aluminum paint on canvas, 87 × 118 (2.210 × 2.997). National Gallery of Art, Washington. Gift of R. Horace Gallatin.*

action cannot be separated from the biography of the artist. The action painting has the same metaphysical substance as the existence of the painter. The new way of painting has made any distinction between art and life superfluous."[14] In other words, the end result of the act of "action painting" is the calligraphic signature of the individual who acted it out.

**Figure II.13** *Jackson Pollock (1912–1956).* Number 5, 1952. *1952. Enamel on canvas, 56″ × 31 1/2″. Collection Lee Krasner Pollock.*

**Figure II.14** *Willem de Kooning (1904– ).* Woman I. *1950–52. Oil on canvas, 6′3⅞″ × 58″. Collection, Museum of Modern Art, New York. Purchase.*

Although Pollock did not work from sketches or studies, but directly from the unconscious, he made an interesting comment about critics labeling his work: "I don't care for 'abstract expressionism,' and it's certainly not 'nonobjective,' and not 'nonrepresentational' either. I'm very representational some of the time, and a little all of the time. But when you're painting out of your unconscious, figures are bound to emerge."[15]

Early on Willem de Kooning had worked with figures and with portraits. He did not restrict himself solely to abstraction. The figure, usually female, was central to much of his work throughout his life. *Woman I*, along with a series of other painted female images swung from sex to glamour to mythological beast (note legs and hoofs on *Woman I*).

But during the middle and late fifties, de Kooning returned to abstraction. *Gotham News* condenses the frenetic energy of New York. His strokes are big and pushy, physically intense, a signature of Abstract Expressionism.

Counter movements began pushing against abstraction by incorporating parodies of well-known works.

In 1976, Mel Ramos poked fun at de Kooning's busty aggressive female, turning her posture, stark glare, and feminine interest to the original painter, Willem, indicated by the title, *I Still Get a Thrill When I See Bill.*

Franz Kline was transformed to his version of action painting when he projected fragments of his small sketches onto a wall and saw " . . . the implications as large-scale, free abstract images."[16]

Steve Martin "appropriates" a Franz Kline by imposing himself before the painting. He is dressed in white tie and tails on which paint is stroked much like the canvas. Martin becomes, in this tongue and cheek performance action piece, the correlative other, the figure. Martin has moved form against figure and poses as a figure against form.

**Figure II.15**   *Willem de Kooning (1904–).* Gotham News. *1955. Oil on canvas, 68″ × 79″. Collection Albright-Knox Art Gallery, Buffalo. Gift of Seymour H. Knox, 1955.*

**Figure II.16**   *Mel Ramos.* I Still Get a Thrill When I See Bill. *1976. Oil on canvas, 80″ × 70″. Inventory #117. Courtesy Louis K. Meisel Gallery, N.Y. Photo by Bruce C. Jones.*

**Figure II.17**   *Steve Martin "appropriating" a Franz Kline. Courtesy Sidney Janis Gallery. Photo: © Annie Liebowitz. Contact Press Images.*

**Figure II.18** *Josef Albers (1888–1976).* Homage to the Square: Apparition. *1959. Oil on board, 47 1/2″ × 47 1/2″. The Solomon R. Guggenheim Museum, New York. Photo: David Heald.*

Josef Albers, a painter and designer with a long and distinguished academic career, began a series of paintings in 1950 titled *Homage to the Square.* His legacy as a geometric abstractionist through his experimentations in economy of form shows squares of color within squares of color. Subtleties of hues in proximity to each other allow his art to distance itself radically from nature. "Art should not represent but present," was his rationale.[17]

Paul Cadmus' beautifully rendered renunciation of geometric abstraction places a nude male inside a quasi-Alberian receding square. The painted square is cracking and peeling at its edges as the male reads a newspaper titled "Nuclear Review." "Art for *your* life-style" is written in the upper right corner of the peeling paint.[18]

How does the reasoning behind this paradigm contribute to figure drawing? Consider again, that if you draw from the figure, in whatever way, you are bound by the figure. If you draw with form as content, in whatever way, you are bound by your own mind and spirit. Are both really antithetical to the other? Does each side have to taunt the other? Maybe, just maybe, an artist can reference a figure better having worked solely in form. And maybe, just maybe, one can reference form better after studying the figure.

Studies from a number of students illustrate my point.

**Figure II.19** *Paul Cadmus.* The House That Jack Built. *1987. Egg tempera on gesso panel, 33 inches square. Courtesy Midtown Payson Galleries, N.Y.C.*

## A Study:

Michael Schonhoff began work in a design class one year prior to figure drawing. Toward the end of his class in design he felt restricted with media and scale, and began working on small pieces of material, using paint, studying its effects. He liked the simple manipulation of the elements, the lines, values, and spaces.

Schonhoff built his large piece by working smaller areas much like the diminutive "sketches" he had done earlier. If he liked an area well enough, he would make the rest of the field around it work. "The piece was finished when I felt good about it, when I found myself fully in it. It was a judgement call."[19]

When asked how his work in abstraction helped resolve his work with the figure, he responded that the figure could be taken apart by each formal element, that the figure was lines, values, colors, textures, shapes, and space.

Schonhoff felt that through the human being, the model (male or female), he could go deeper into himself for the following reasons. He could use the figure as a gestural springboard for a series of studies. By bringing chance markings into the work, he could build up space all over his drawing paper, just as he had in his action painting. Moving back and forth between the biographical marks and the model's body attitude, he worked "personal colors" into the image, eventually calling his process a "dance of the mind."

Schonhoff cannot sit down and draw. His process is fast. The more his work moves, the more he moves. He is not at ease doing the work. The whole piece is a struggle; sometimes stopping is difficult, unless he is interrupted.

**Figure II.21**  *Michael Schonhoff.* Figure. *1990. Grease base crayons on rag paper, 30″ × 24″.*

**Figure II.20**  *Michael Schonhoff.* Untitled Objective. *Fall, 1989. Acrylic, ink, and charcoal on canvas, 8′ × 4′. Private Collection.*

**Figure II.22**  *Michael Schonhoff.* Two Figures. *1990. Grease base crayons on rag paper, 30″ × 24″.*

## A Study:

Katie Stone, on the other hand, started with the usual fare for figure drawing, i.e., gestures, sighting, volumes, and anatomy. Gradually she began working the negative areas the body implied. Her scale initially was small, intimate. Soon the scale became larger, the figures more sweeping, more suggestive, more abstract.

Over a period of one year Stone worked in a painting studio using large plastic sheets as a backdrop to cover the wall. Each work she finished on paper, when removed from the wall, left a residual shape on the plastic sheet much like a stencil. Gradually Stone became more engaged with the "wall remnant shapes" than she did with the figure, until her use of rectangles and writing superseded her interest in the figure. Her figures seemed, for her, redundant shapes.

Stone began the two following works in this way. She used smaller sheets of drawing paper as her stencils, to begin with some structure in the work, suggesting a sense of shields. She wrote on the canvas with graphite stick—a dream, thoughts that did not make sense. She wanted to build a surface, a texture. Then with many thin layers of paint, using colors as values, she pushed and pulled the plasticity of the surface, until she felt the work was done.

*Figure II.24*    *Katie Stone*. Shield II. *1990. Acrylic on paper, 40″ × 29″.*

*Figure II.23*    *Katie Stone*. Birth. *1989. Acrylic on rag paper, 22″ × 30″.*

*Figure II.25*    *Katie Stone*. Shield I. *1990. Acrylic on Paper, 40″ × 29″.*

## A Study:

Another student, Donna Skripsky, comes to her finished works by way of her thoughts about people or her reactions to the lyrics of music and song. She begins layering colors, in this case, turquoises and reds, mixed with modeling paste as an extender. With a big brush, a trowel, or a knife, she spreads the colors onto large sheets of paper, creating a texture.

Skripsky then applies rice or mulberry papers with matte medium, layering as needed. She paints on the smaller paper shapes after they adhere to the surface. Finally, large areas of black are worked into spaces in the painting until she feels the work has a unified motion.

## A Study:

Adrian Penn went from figurative work to abstractions because he " . . . was not comfortable with the figure."[20] Many of his early works show portions of figures, rarely the full figure, because he liked to infer mystery. That same sense of mystery declares itself in "Self Portrait."

While Penn was in the library, he picked up a music analysis book, found a Mozart score and photo-copied it.

Apart from enjoying Mozart, Penn liked the score as an abstract design. He began building his painting off that shape, the rectangle. He overlaid a couple of pieces of loose canvas onto his stretched canvas to continue building his composition. Two other photocopies, this time the trajectories of military air-to-air missile charts, helped to bring elements together. Imposed in the lower right is a series of rusty nails, leftovers from other works reminiscent of his stylized trees of earlier drawings. The acrylic paints full of reds, browns, burnt and raw umbers, oxides, and raw siennas were moved and pushed for values, shapes, space, and color.

## A Study:

Diane Gail begins her work from the male or female model in the classroom. She draws one pose at a time. As the model moves to a new pose, Gail turns her paper 90 degrees and draws over the prior image. She continues drawing with litho crayon until her paper is covered but not totally blackened.

Later, at home, Gail pours paint thinner over the drawing allowing it to "set" for a short time. The thinner softens the litho so that the drawing can be used as a printing plate.

**Figure II.26** *Donna Skripsky.* DS 490, 1990. *1990. Acrylic on rag paper, 46″ × 39″.*

**Figure II.27** *Adrian Penn.* Self-Portrait. *1990. Acrylic on canvas, 4′ × 3′.*

Gail takes a clean, second sheet of paper and presses it on top of the thinner-soaked one. She moves the sheets to blur the drawn figurative forms and to suggest movement in the new composition. The process is repeated as many times as is necessary to achieve the spatial effects she is after, sometimes adding more litho to the original wet surface. Once the applied sheet has the characteristics of the space Gail wants, she stops the mono-printing process and lets the newly printed sheet of drawing paper dry. Sometimes she moves back into the piece with her litho crayon to complete the abstraction by increasing varieties of depth.

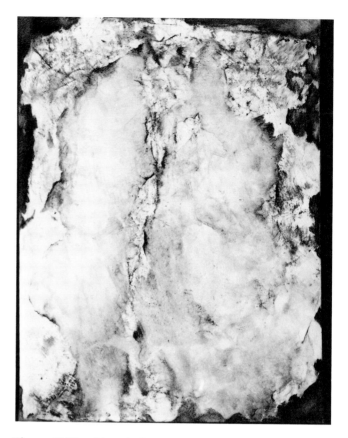

***Figure II.29***   *Diane Gail.* No. Two. *1990. Litho crayon, thinner, mono-print on rag paper, 24" × 18".*

***Figure II.28***   *Diane Gail.* No. One. *1990. Litho crayon, thinner, mono-print on rag paper, 24" × 18".*

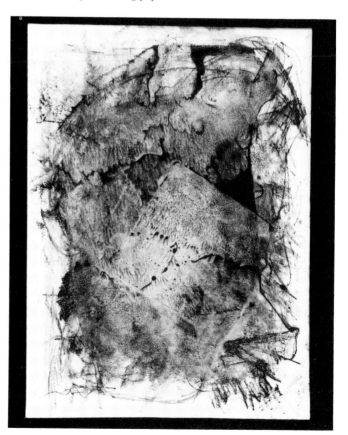

***Figure II.30***   *Diane Gail.* No. Three. *1990. Litho crayon, thinner, mono-print on rag paper, 24" × 18".*

## A Study:

Tom Hansen's trilogy was selected from a group of ten works. He began with random graphite markings on rag papers about 20″ high by 17″ wide. As Hansen kept going, figurative images surfaced without his making conscious decisions. Because the scribble marks originated from the subconscious and the layers of markings produced uncontrolled images, Hansen feels " . . . these figurative works become a new way of imaging more of a psychological approach, instead of reproducing deadly anatomical renderings."[21]

**Figure II.32**    *Tom Hansen.* Recognition II. *1990. Graphite on rag paper, 20″ × 17″.*

**Figure II.31**    *Tom Hansen.* Recognition I. *1990. Graphite on rag paper, 20″ × 17″.*

**Figure II.33**    *Tom Hansen.* Recognition III. *1990. Graphite on rag paper, 20″ × 17″.*

## A Study:

Derek Anderson is another student who makes marks on his drawing papers, sometimes referring to a model, sometimes not. Smearing the oil base and the colored crayon marks (he uses four colors per drawing as a picture base) with an eraser, he rotates his paper a number of times until an image surfaces. Once an image does appear, Anderson moves back into the piece, reworking sections, to solve problems with composition, color, texture or borders. When the surface is too chaotic to build any more, he quits. And always there is wit in Anderson's work. His high intensity reds, greens, blues, yellows, oranges, and purples produce *non sequitur* images that finally make sense in their non-sense.

## A Study:

Don Rouse also uses the method described above, incorporating a darker palette of hues in oil-based crayons. The images that surface for him are " . . . intended to illustrate the eerie and uncomfortable feelings of the male/female struggle for control in a relationship."

**Figure II.35** *Derek Anderson.* Winged Thought. *1990. Oil-based crayons on rag paper, 24″ × 18″.*

**Figure II.34** *Derek Anderson.* Expressions on the Wharf. *1990. Oil-based crayons on rag paper, 24″ × 18″.*

**Figure II.36** *Don Rouse.* Control. *1990. Oil-based crayons on rag paper, 24″ × 18″.*

## A Study:

Jennifer Handevidt begins her work on 30″ × 20″ rag paper. She tapes other pieces of paper alongside the edges of the rectangular shape she has drawn on her original sheet. With paper taped alongside the drawn rectangle's edges, Handevidt is free to build her surfaces without inadvertently moving into the white border field her works need.

Handevidt draws from her life experiences and from her responses to them. She uses deep, brilliant reds, golds, purples, greens, and turquoise. Layer after layer is pushed, rubbed, and subdued with blacks until the work feels finished.

## A Study:

Kris Lucas' imagery in her diptych came out of her mystical forest of people, where she allowed Cubism to play a role. Largely through the use of lines and values she pulls space from the background directly into portions of foreground images. She did a number of drawings using her mythical people and animals, this time hinging two large masonite boards that close and open, much like switches, turning on the imagination or turning it off. Her mixed media of pastels, crayons and graphite indicate a child's world fraught with fears, confronted by a tormentor a child was forced to respect.

## A Study:

Barbara Walton turns to images of Mother Earth. There was no actual figurative reference, just an idea in her head. The materials Walton uses are nontraditional, such as scraps of torn mat board and paper toweling. From those materials she creates the basic composition, after which she used light washes of watercolor, in blues, yellow oxides, reds, and indigos. Gestural lines with a colored pencil are added last. The arrows indicate the imposition of human beings on nature, the purposive tearing of Mother Earth.

**Figure II.38**  *Kris Lucas.* Childhood. *1988. Mixed media on masonite (with drilled holes through which heavy yarn was wound), each panel 56″ × 24″.*

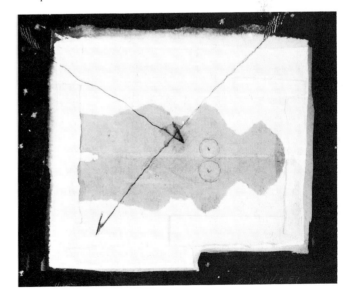

**Figure II.37**  *Jennifer Handevidt.* I Don't Think I Love You Anymore. *1989. Grease crayon, charcoal, and pencil, 30″ × 20″.*

**Figure II.39**  *Barbara Walton.* Hide. *1990. Mixed media, colored pencil, and watercolor, 10″ × 12″.*

169

## A Study:

Peter Hansen's *Irony of the Bored* began with found objects. Hansen had picked up a broken and discarded street barricade, which he kept in his studio for five or six months. Meanwhile, he found an ironing board from the 1940s in his father's attic. Hansen stripped it of layers of soiled ironing cloth. On the ironing board itself was the original brand label, "Rid-jid Ironing Board—will not wiggle, wobble, jiggle, slip or slide."

As time went on, Hansen found the idea for his work also included the barricade pieces. Hansen saw his work as a secular triptych principled on early religious triptychs. He worked some preliminary sketches, each about five minutes. He kept the label on the ironing board but drew a "Medusa Madonna" around the label, outfitting her in a black bra and a back pack, snake-like curlers, complete with halo.

Marilyn Propp, a professional artist living in Chicago, speaks about her work: ". . . from 1979–1986 I worked pastel studies of urban and rural landscapes, which gave me an awareness of color informed by outdoor light . . . That sense of . . . richly glowing colors, has entered the personal world of myth and story. My work has been, and is, constantly informed by both a turning outward and a turning inward. I paint interior dramas peopled by the archetypes and symbols which I encounter daily, both in the world around me and in the world of my interior journey . . . The figures are often life-sized, and are pressed close to the picture plane in a short but not flat space, which creates a further play between surface tension, flatness and spatial concerns . . . figures and objects are cropped, adding to the compression of space and giving the sensation that they are continuing into the viewer's space . . ."[22]

Figuration since the grand sweep of Abstract Expressionism in the 1950s covers many types and styles in as many locales. The historian/author Edward Lucie-Smith in his book, *American Art Now*, says there are two different labels for many American figurative artists " . . . 'New Image' and 'Bad Painting.' "[23]

**Figure II.40** *Peter Hansen.* The Irony of the Bored. *1989. Mixed media (ironing board, barricade, electrical plugs, acrylic paint), 54″ × 30″ (open).*

**Figure II.41** *Marilyn Propp.* Tablets by the Sea. *1990. Pastel, ink, and paper, 12 1/8″ × 12″. Collection of the artist.*

Rather than sort through a network of current artistic 'isms' beyond the boundaries of this text, we will consider some artists who have maintained their own directions for decades, a group we will call *Mainstream Outsiders*. Later we will look at a few artists who have shifted from the roots of abstraction into figurative images, calling them *Transformers*.

Among the "mainstream outsiders" are Alice Neel, Alfonso Ossorio, Leon Golub, and Red Grooms.

Alice Neel described herself as . . . "an old-fashioned painter. I do country scenes, city scenes, portraits and still life."[24] Throughout her life, Neel was one of the most honest, eloquent, and penetrating portrait painters of the twentieth century. About Andy Warhol she said, ". . . He took that shirt off. Now the reason he wears that corset is when the doctors took the bullets out that Valerie Solanis shot him with, they had to cut his stomach muscles. The Africans wouldn't let me show it for my slide lecture on TV that I gave in Nairobi, Africa. They said: 'Oh, no, half man, half woman, it would ruin our children.' "[25]

Alfonso Ossorio, a Philippino, was associated with early Abstract Expressionism, but even in 1949 produced *Birth*, a work of grotesque art. Robert Doty, Curator of the Whitney Museum of American Art in 1969, defined the "grotesque" as an art form in itself for these reasons: reason was rejected, the subconscious was plunged, opposites were visible (i.e., the grotesque with the beautiful), and there was emphasis on virulence, ridicule, or surprise in deformation and caricature.[26]

*Figure II.42* *Alice Neel (1900–1984). Andy Warhol. 1970. Oil on canvas, 60″ × 40″ (152.4 × 101.6 cm). Collection of Whitney Museum of American Art, New York. Gift of Timothy Collins. 80.52*

*Figure II.43* *Alfonso Ossorio. Birth. 1949. Wax, watercolor on paper, 40″ × 30″ Collection Helen Harding. Estate of Alfonso Ossorio. Photographed by Eric Pollitzer.*

**Figure II.44**   *Leon Golub.* Wounded Sphinx. *1988. Acrylic on canvas, 120″ × 154″. Josh Baer Gallery, N.Y.C.*

Leon Golub has long been known for his huge, confrontational, *mercenary* political works, but he has moved more recently to archetypal imagery, still incorporating the grating beauty found in his earlier aggravated and scraped surfaces. This time he uses the man/beast sphinx as symbol. Myth is closer to life than one might expect. Often in Golub's *mercenary* paintings were strained body postures, brilliant blood reds, and grated surfaces with a heightened sense of confinement. In these sphinx paintings, as Robert Storr suggests in his article *Riddled Sphinxes,* the protagonists appear cornered.[27] Perhaps for Golub the human condition is one of being cornered, of being hurt, confused, or bound.

Red Grooms, a gifted artist of humorous images comments on his work of the 1960s: "I wanted to be a contrast to the heavy esthetic side of art. There was so much formalism in art that I felt like LeRoy Neiman or something."[28]

His documentaries—watching what people wear and do—are described in one instance by Minnesota master printer, Steven M. Anderson: "Once when Red was here, he needed to send a card to some friends back in New York, so he left the studio for a few minutes to drop into a card shop he'd seen down the street. Well, a few minutes later he came running back in, shouting, "Quick, where's something I can draw on?" There were some lithographic stones laid out on the counter, and he began to draw right onto one of them a caricature of this new-wave girl who was the clerk at the card shop. She had purple hair and sunglasses, and he put her in this leopard-skin outfit. Then behind her he added some silos and put the reflection of a truck in her sunglasses. She was supposed to be the Minneapolis version of "Christina's World" . . . He couldn't think of a name for her, though, so he went back to the card shop and asked what her name was. It was Lorna Doone. That name absolutely made Red's day—his week."[29]

*Figure II.45* Red Grooms. Lorna Doone. *1979–1980. Color lithograph with collage and rubber-stamp impressions printed on 2 sheets 24 1/2″ × 32″ each. Edition of 48. Collection Brooke Alexander Gallery. Photographed by Eera-inkeri Inc., New York, NY.*

*Figure II.46* Ed Paschke. Lucy. *1973. Oil on canvas, 60¼″ × 38″ (153.0 × 96.5 cm). Museum of Contemporary Art, Chicago, Gift of Albert J. Bildner.*

The *Mainstream Outsiders* painted to the "beat of their own vision," regardless of other artistic awakenings.

Most of the following artists, The Transformers, were bred on Abstract Expressionism. Most have lived through proponents of that school with teachers, peers, museums, and galleries. Such exposure was bound to have an effect on their figurative work.

H. H. Arnason says this about the generations following Abstract Expressionism: "Despite the general sense that Post-Modernism means revived figuration, the New Image and Neo-Expressionist painters, for the most part, treat the human figure as incidental to some overriding interest in problems of form, process, expression, or didactic theme. Few address their art totally to the figure in all its naked physical and psychological fullness, especially as this can be revealed in ambiguous but recognizably banal relationships among two or more images."[30]

Ed Paschke, Jim Nutt, and Gladys Nilsson were New Imagists, Chicago style. "In sum, Imagism, the prevailing Chicago esthetic—a Surrealist-influenced, highly personal, and expressive attitude in which figurative imagery dominated—was clearly in place by the late 1950s."[31]

Ed Paschke absorbed Marshall McLuan's adage, "The Medium is the Message." Paschke is quoted, "My belief is that these elements (such as television), for better or worse, have woven their way into the collective fabric of our lives. For me, the distinction between direct experiences and those which are modified through mass media is becoming smaller and smaller."[32]

His hermaphroditic *Lucy* of 1973 is not erotic nor sensual, but plausible as an intractable composite figurative clue to human nature.

The humanoid, mutilated women of Jim Nutt's early work in the 60s has moved to Cubistically-stylized portraits, still pursuing a psychosexual male/female mythological drama. His figures seem emblematic of human relationships.

Gladys Nilsson's *Pandemoneeum—A Trip-Dick,* is a Hieronymous Bosch-ian Javanese puppet show. Our eye level is closer to the ground than in Bosch's "Garden of Delights," but a dance of sexual provocation permeates the stylized atmosphere with tongue-in-cheek humor (pun intended).

Turning briefly to New York, Jean-Michel Basquiat, a self-taught graffiti artist born in the 1960s, has now exchanged aerosols and magic markers for paint as a medium with which to draw. The strokings of this "primitive" has a pervasive rigour and toughness one associates with "street smarts." Basquiat has an " . . . unteachable sense of how to balance the contradictory forces of his primitivism and sophistication, immediacy and control, wit and savagery."[33]

*Figure II.47*   *Jim Nutt.* Proof (proof). *1987. Acrylic on masonite and wood, 27⁷⁄₁₆ × 22⅜. Photo courtesy Phyllis Kind Galleries, Chicago and New York. Photo credit William H. Bengtson. Private collection.*

*Figure II.48*   *Gladys Nilsson.* Pandemoneeum—A Trip-Dick. *1983. Watercolour on paper 51″ × 101¼″. (129.5 × 257 cm). Photo courtesy Phyllis Kind Galleries, Chicago and New York. Photo credit William H. Bengtson. Private collection.*

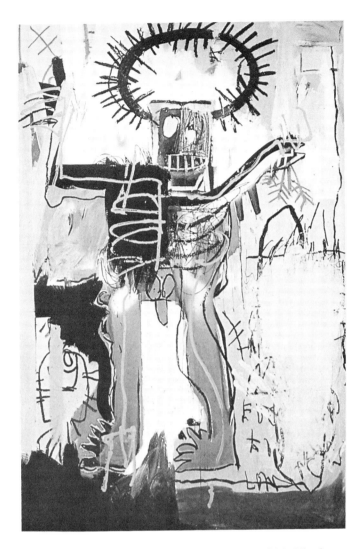

**Figure II.50** *Luis Cruz Azaceta (Cuban, b. 1942).* Homo Fragile. *1983. Acrylic on canvas, 72¼″ × 120¼″. (182.5 × 304 cm). Archer M. Huntington Art Gallery, The University of Texas at Austin, Archer M. Huntington Museum Fund, 1987. Photo credit: George Holmes.*

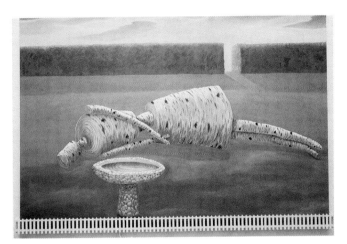

**Figure II.49** *Jean-Michel Basquiat.* Untitled. *1981. Mixed media on wood panel, 73 1/4″ × 49 1/4″. Collection Robert Lehrman, Washington, D. C. [Seen in* Arts Magazine, *Vol. 64, No. 7, February 1990. Illustration on page 55.]*

**Figure II.51** *Cheryl Laemmle.* Empty Birdbath. *1983. Oil on canvas, 56″ × 84″. (142 × 213 cm). Courtesy Sharpe Gallery, New York [Seen in* American Art Now *by Edward Lucie-Smith, Phaidon Press, Oxford, 1985. Illustration 105 on page 66.]*

The Cuban-born, New York resident, Luis Cruz Azaceta painted a work titled *Homo Fragile* in 1983. His explosive stroke dynamics are reminiscent of some abstract expressionism. Azaceta emigrated in 1960 to New York and did a stint in the subways in 1974, developing his iconography. "I create certain fantasies in order to make a point, to raise consciousness. Sometimes the work is shocking, but never more shocking than reality. Reality is pretty brutal . . . in New York you have extremes. There's both the really good and the really bad . . . New York has the intensity of extremes. I try to depict the victim, usually, not the aggressor. It's hard to say sometimes who the aggressor is. But I know who the victim is."[34]

Cheryl Laemmle uses provocative figurative forms that bear an anthropomorphic resemblance to body attitudes of human beings, with birchwood shapes. A throwback to the early twentieth century metaphysical, de Chirico, is suggested by the manikin figure in the human-made idyllic environment.

Jonathan Borofsky has a trademark for installations: ". . . the side of me which is the core of feeling—the heart, the spirit—is connected with the conceptual side of me . . . I've put the two together, physically and symbolically."[35] Alternative realities are confronted when one enters a Borofsky show. So much is going on—movements, sound, visual contradictions on floor, wall, and

*Figure II.52* *Jonathan Borofsky exhibition installation, Paula Cooper Gallery, New York, 1983. Courtesy, Paula Cooper Gallery. Photo by Geoffrey Clements.*

ceiling—that one is tempted to feel a part of an action painting, as though we were small enough to be just beneath the drips of Pollock's action painting.

Charles Garabedian, a California artist, paints a neo-Ulysses who floats between reality and myth. His figure is both classical and primitive, yet it embodies the abandonment and traumatic emotional content of the late twentieth century.[36] This work evidences a distorted field of illusion. The bow of the boat carries a limited illusion of perspective and only a few areas indicate cast shadows. Most of all, the entrance into the work and movement within the composition are vertical and flat. Horizontal repetitions continue to move the eye upward in the basic abstractions of the composition construction. This piece contains an odd collection of items, including the naked Ulysses—disjunctive, removed, solitary, sad.

Turning now toward the figurative realists, we will look at six artists who work in very different ways.

In the 1950s and 1960s, Gregory Gillespie began making realistic art, and has continued since, even though he does not like the label. His work is meticulous in detail. Often using himself ("he's always around," he says) as a

*Figure II.53* *Charles Garabedian. Ulysses. 1984. Acrylic on canvas, 90″ × 66″. The Eli Broad Family Foundation, Santa Monica, Calif.*

176

*Figure II.54* *Gregory Gillespie*. Myself Painting a Self Portrait. *1980. Mixed media on panel, 58 1/2″ × 68 3/4″ Photo courtesy Forum Gallery.*

model, his work, *Myself Painting a Self Portrait,* Gillespie reveals a confident humor many artists do not betray. The portrait his own painted image is painting is smiling as broadly as the famous Cheshire Cat of Lewis Carroll's *Alice in Wonderland.*

Erich Fischl's work often makes the viewer an unwitting voyeur participant. His people are frequently ambiguous and carry a sinister or sexual mystery about them because of what they are doing or what they are carrying, or by the titles that infer something about them other than what the image itself might suggest.

For instance, his title *Imitating the Dog—(Mother and Daughter II),* conjures up all kinds of references. In another work, *Girl with Doll,* Fischl paints a wiry, knobby-kneed, nude prepubescent girl standing outside, holding a large stuffed monkey puppet with arms clinging to her neck. A naked eleven-year-old girl in the sunlight with a hairy monkey dangling around her body makes eerie nonsense.

Our illustration here titled, *Haircut,* infers an upper-middle-class ritual bathroom scene. We are uneasily a voyeur to something about to happen.

*Figure II.55* *Eric Fischl.* Haircut. *1985. Oil on linen, 104″ × 84″. The Eli Broad Family Foundation, Santa Monica, California. Photographed by William Nettles.*

***Figure II.56*** *William Beckman.* Diana IV. *1981. Oil on wood panel, 84 1/2″ × 50 7/8″. Hirshhorn Museum and Sculpture Garden, Smithsonian Insitution, Washington, D.C. The Thomas M. Evans, Jerome L. Green, Joseph H. Hirshhorn, & Sydney & Frances Lewis Purchase Fund.*

***Figure II.57*** *Audrey Flack.* Marilyn (Vanitas). *1977. Oil over acrylic on canvas, 96″ × 96″. Collection of the artist. Terry Allan. Photo courtesy Louis K. Meisel Gallery, New York. Photo credit: Bruce C. Jones.*

William Beckman, on the other hand, works with a delicacy of stroke that moves almost into photo-realism. You might remember that artists of early collages used materials that served two functions: one was to be a part of the work itself, the second was to present the materials simply as materials. Of Beckman's work, few accoutrements detract from the personality of the model. *She* becomes a presence just as much as the painting does.

Audrey Flack, an air brush photo-realist, is inspired by the over-stated and luxurious Baroque traditions in art. She uses a camera to reach for a more intense and personal involvement with her subject matter. In the late 1950s and the 1960s, when Flack wanted to move away from a Cézannesque influence, she shifted to photographic portraiture. In 1971 she began a series of works that were "made possible by the convergence of her interest in art and photography, her historicism, and her sympathy for the Baroque."[37]

Audrey Flack says of her portrait of Marilyn Monroe, " . . . I chose a photograph of Marilyn that shows her character in transition. Her face retains qualities of Norma Jean and has not yet fully become Marilyn Monroe. Her hair is still soft—it curled and flowed along with my air brush . . . the mouth is becoming plastic but has not yet firmly set, although the lips are beginning to crack. . . ."[38]

Martha Meyer Erlebacher's early work in the 1960s rested in optical phenomena. After beginning to work with subject matter, Erlebacher drew from life at every opportunity, feeling that the figures she wanted to draw had to be drawn correctly. Admonitions aside—"do not draw realistically" (heard so often in art schools in the 1950s and 1960s)—she moved on with her recorded drawings because she wanted "structure" in her figures. And she has studied the figure to such length that she can have it stand or move as she sees fit in her imaginative scenes. The figures are realistic, but the settings sometimes are anachronistic, sometimes a backdrop for anatomy with a twist.

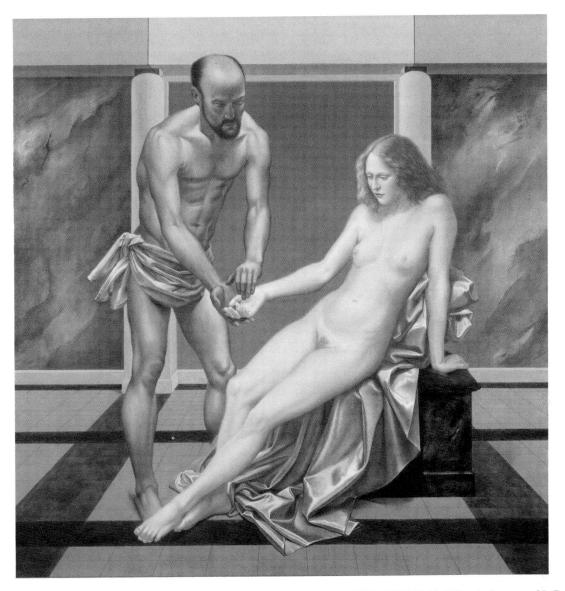

**Figure II.58** *Martha Mayer Erlebacher*. Mars and Venus. *1983. Oil on canvas, 52″ × 52″ (132 × 132 cm). Courtesy of J. Rosenthal Fine Arts, Ltd. Chicago, IL. [Seen in* American Art Now, *by Edward Lucie-Smith. Phaidon Press, Oxford, 1985. Illustration No. 184 on page 103.]*

Our illustration, *Mars and Venus*, has classical subject matter and poses, but the male and female are contemporary, the bodies are not ideal. They are placed in a "Magritte-ian" surrealistic space, a curious, odd suspension of belief.

Jennifer Bartlett's huge, 153′, room-consuming *Rhapsody* is not a work about figures so much as it is a work that takes us back through the hollow echoes of the mind to Cézanne, which is about where we began this paradigm.

*Rhapsody*, according to Roberta Smith, " . . . is a great and imperfect epic, a visual event that unfolds and refolds in real time, real space, and above all, real thought, without ever leaving the wall."[39]

The piece is like music and literature via the paintbrush, a piece with themes, cadences, motifs, and resolutions. The work is made up of plates, each about page size. Bartlett begins with Minimalism and moves to some childish line drawings ("bad painting"). Cycles of representation and abstraction elicit satiric throwbacks to other artists and methodologies.

The work begins like building blocks in vertical rows with seven units in each—solid color units, a mountain, more color plates, a house, two trees, color plates again, horizontal lines, and so on. Sometimes the natural world steals in, eroding abstractions and vice versa.

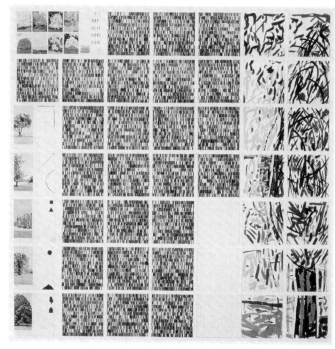

(a.)

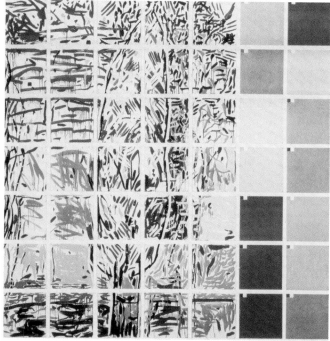

(b.)

***Figure II.59 a, b, c*** *Jennifer Bartlett.* Rhapsody. *1976. 987 one-foot-square steel plates baked with white enamel, silkscreened with a gray quarter-inch grid then painted on with enamel and a small brush, arranged in 142 rows with about 7 plates each. The work needs 153 running feet of wall for proper installation. [Seen in* Rhapsody—Jennifer Bartlett. *Introduction by Roberta Smith with notes by the artist and photographs by Geoffrey Clements. Harry N. Abrams, Inc., Publishers, New York, 1985. Illustrations excerpted from pages 39, 40, and 41 (rows 85 through 105).]*

(c.)

Our illustration shows the square units in rows 84 through 105, with trees in full bloom, moving to vertical color strokes that create a field of brush strokes paradoxically both broken and bound by the separated units. Then, the color strokes are loosed until Bartlett moves to square units of solid colors, then to black and white geometric shapes that play a hybrid leap-frog with Cézanne.

As we end this paradigm and move to the next—*Figure Above Form*—which explores both cultural and religious icons, remarking on one of Norman Rockwell's famous cultural icons, *The Connoisseur*, might be in order. The studious, open-for-any-information-please stance of the middle-aged and perplexed "everyman" appears to be asking of the Abstract Expressionist Jackson Pollock, "Is

**Figure II.60** *Norman Rockwell (1894–1978).* The Connoisseur. *Original oil painting for a* Saturday Evening Post *cover January 13, 1962. Photo courtesy of the Normal Rockwell Museum at Stockbridge, MA. [Seen in* Norman Rockwell's People *by Susan E. Meyer. Harry N. Abrams, Inc., Publishers, New York, 1981. Illustration on page 18.]*

this art?" "What does it mean?" "Why is this thing in a museum?" "Why should I care?" "What am I missing?"

## ENDNOTES

1. Edward Lucie-Smith, 1984. *The Thames and Hudson Dictionary of Art Terms.* London: Thames and Hudson, Ltd., p. 102.

2. Arsen Pohribny. 1978. *Abstract Painting.* Oxford: Phaidon Press Limited, p. 13.

3. H. W. Janson. 1969. *History of Art.* Revised Ed. Englewood Cliffs, N.J.: Prentice-Hall, Inc., and New York: Harry N. Abrams, Inc., p. 507.

4. Ibid, p. 527.

5. Pohribny. *Abstract Painting.* p. 51.

6. H. H. Arnason. *History of Modern Art.* Third Ed. Revised and updated by Daniel Wheeler. Englewood Cliffs, N.J.: Prentice-Hall, Inc., and New York: Harry N. Abrams, Inc., p. 127.

7. Ibid, p. 187.

8. Ibid, p. 187.

9. Ibid, p. 203.

10. Pohribny. *Abstract Painting.* pp. 11–12.

11. Harold Osborne. Ed. 1970. *The Oxford Companion to Art.* Oxford: Clarendon Press, p. 395.

12. Arnason. *History of Modern Art.* p. 387.

13. Ibid, p. 388.

14. Pohribny. *Abstract Painting.* p. 30.

15. Selden Rodman. 1961. *Conversations with Artists.* New York: Capricorn Books, p. 82.

16. Arnason, *History of Modern Art.* p. 391.

17. Harold Osborne. Ed. 1988. *The Oxford Companion to Twentieth-Century Art.* Oxford: Oxford University Press, p. 12.

18. Jonathan Weinberg. March, 1988. "Paul Cadmus at Midtown." *Art in America,* V. 76. p. 147.

19. An interview with Michael Schonhoff, December, 1990.

20. An interview with Adrian Penn, Fall of 1990.

21. An interview with Tom Hansen, Fall, 1990.

22. From *Artist's Statement* by Marilyn Propp, 1990.

23. Edward Lucie-Smith. 1985. *American Art Now.* Oxford: Phaidon Press Limited, p. 29.

24. Ibid, p. 19.

25. Patricia Hills. 1983. *Alice Neel.* New York: Harry N. Abrams, Inc., p. 138.

26. Robert Doty. 1969. Curator, Whitney Museum of American Art. *Human Concern/Personal Torment: The Grotesque in American Art.* New York: Praeger Publishers. No pages listed.

27. Robert Storr. March, 1989. "Riddled Sphinxes." *Art in America.* V. 77, pp. 126–131.

28. *Red Grooms, A Retrospective 1956–1984. 1985.* Essays by Judith Stein, John Ashbery and Janet Cutler. Pennsylvania Academy of the Fine Arts, p. 13.

29. Ibid, p. 12.

30. Arnason. *History of Modern Art.* p. 647.

31. Neal Benezra. 1990. *Ed Paschke.* New York: Hudson Hills Press for The Art Institute of Chicago, p. 19.

32. Ibid, p. 35.

33. Arnason. *History of Modern Art.* p. 655.

34. Charles Hagen. February, 1990. *Art News.* Vol 89, No. 2., p. 117.

35. Sam Hunter. Ed. 1986. *An American Renaissance Painting and Sculpture Since 1940.* Museum of Art, Fort Lauderdale, Florida. New York: Abbeville Press Publishers, pp. 184–185.

36. Ibid, p. 184.

37. *Audrey Flack on Painting,* notes by Ann Sutherland Harris, introduction by Lawrence Alloway, New York: Harry N. Abrams, Inc., 1985, p. 24.

38. Ibid, p. 84.

39. Roberta Smith. *Rhapsody, Jennifer Bartlett.* New York: Harry N. Abrams, Inc., p. 7.

# BIBLIOGRAPHY

Arnason, H. H. *History of Modern Art.* Third Ed. Revised and updated by Daniel Wheeler. Englewood Cliffs, N.J.: Prentice-Hall, Inc., and New York: Harry N. Abrams, 1986.

Aronson, Steven M. L. "Portrait of the Artists: Eric Fischl and April Gornik on Long Island." *Architectural Digest, 46* April, 1989.

————. *Audrey Flack on Painting,* notes by Ann Sutherland Harris, introduction by Lawrence Alloway, New York: Harry N. Abrams, Inc., 1981.

Auping, Michael. *Abstract Expressionism: The Critical Developments.* New York: Harry N. Abrams, Inc., 1987.

Auping, Michael. *Abstraction-Geometry-Painting, Selected Geometric Abstract Painting in America Since 1945.* New York: Harry N. Abrams, Inc., In Association with Albright-Knox Art Gallery, 1989.

Benezra, Neal. *Ed Paschke.* With contributions by Dennis Adrian, Carol Schreiber, and John Yau. Hudson Hills Press, New York, 1990.

Berman, Avis. "Artist's Dialogue: Eric Fischl, Troubles in Paradise." *Architectural Digest, 42,* no. 72 December, 1985.

Boone, Mary. "Eric Fischl." *Art News, 87,* September, 1988.

Doty, Robert, Curator, Whitney Museum of American Art, *Human Concern/Personal Torment, The Grotesque in American Art.* New York: Praeger Publishers, 1969.

Frank, Elizabeth. *Jackson Pollock.* New York: Abbeyville Press, 1983.

Frank, Peter and McKenzie, Michael. *New, Used & Improved Art for the '80s.* New York: Abbeville Press, 1987.

Franke, Herbert W. *Computer Graphics-Computer Art.* New York: Phaidon Press Limited, 1971.

Friedman, B. H. *Alfonso Ossorio.* New York: Harry N. Abrams, Inc., 1972.

Gaugh, Harry F. *Willem deKooning: Modern Masters Series.* Abbeville Press, Inc., 1983.

Gilmour, Pat. *The Mechanised Image, An Historical Perspective On 20th Century Press.* Arts Council of Great Britain, 1978. CTD Limited, Twickenham, England.

*Red Grooms, A Retrospective, 1956–1984.* An illustrated catalogue with essays by John Ashbery, Janet K. Cutler, and Judith Stein. Philadelphia: Pennsylvania Academy of the Fine Arts, 1985.

Haftmann, Werner. *Painting in the Twentieth Century, An Analysis of the Artists and Their Work, Volume I.* New York: Holt, Rinehart and Winston, 1965.

Haftmann, Werner. *Painting in the Twentieth Century, A Pictorial Survey, Volume II.* New York: Holt, Rinehart and Winston, 1965.

Hills, Patricia. *Alice Neel.* New York: Harry N. Abrams Inc., 1983.

Hills, Patricia and Tarbell, Roberta K. *The Figurative Tradition and the Whitney Museum of American Art (Paintings and Sculpture from the Permanent Collection).* Newark: University of Delaware Press, 1980.

Hopkins, Henry T. *California Painters: New Work.* San Francisco: Chronicle Books, 1989.

Hunter, Sam; Essay editor and introduction. *An American Renaissance, Painting and Sculpture Since 1940.* New York: Abbeville Press Publishers, 1986.

Jacobs, Michael. *Nude Painting.* Oxford: Phaidon Press Limited, 1979.

Janson, H. W. and Janson, Dora. *History of Art.* Englewood Cliffs, N.J.: Prentice-Hall, Inc., and New York: Harry N. Abrams, Inc., 1964.

Johnson, Ken. "Jim Nutt at Phyllis Kind." *Art in America, 76,* (December, 1988).

Kitson, Michael. *The Age of Baroque.* London: Published by Paul Hamlyn Limited, 1966.

Kuspit, Donald. *Leon Golub, Existential/Activist Painter.* New Brunswick, N.J.: Rutgers University Press, 1985.

Larson, Kay. "An Artist's Sense and Sensibility; Audrey Flack in Manhattan and East Hampton."*Art In America. 7,* January, 1989.

Lerner, Abram. *Gregory Gillespie.* Washington: Published by the Smithsonian Institution Press, 1977.

Lovejoy, Margot. *Postmodern Currents, Art and Artists in the Age of Electronic Media.* Ann Arbor, Mich.: UMI Research Press, 1989.

Lucie-Smith, Edward. *Art in the Seventies.* Oxford: Phaidon Press Limited, 1980.

Lucie-Smith, Edward. *American Art Now.* Oxford: Phaidon Press Limited, 1985.

Magnenat-Thalmann, Nadia and Thalmann, Daniel. *Computer Animation; Theory and Practice.* New York: Springer-Verlag and Tokyo: Sanshodo Printing, 1985.

Marshall, Richard and Rosenthal, Mark. *Jonathan Borofsky.* New York: Harry N. Abrams, Inc., 1984.

Meyer, Susan E. *Norman Rockwell's People.* New York: Harry N. Abrams, Inc., 1981.

Nairne, Sandy. *Women's Images of Men.* London: Institute of Contemporary Arts, 1980.

Pohribny, Arsen. *Abstract Painting.* Oxford: Phaidon Press Limited, 1979.

Read, Herbert. *Concise History of Modern Painting.* New York and Washington: Frederick A. Praeger, Publishers, 1959.

Reid, Calvin. "Kind of Blue." *Arts Magazine, 64,* no. 7, February, 1990.

Rose, Bernice. *Drawing Now.* The Museum of Modern Art, New York, 1976.

Sandler, Irving. *The New York School: The Painters and Sculptors of the Fifties.* Icon Editions, New York: Harper and Row, 1978.

Smith, Roberta. *Rhapsody—Jennifer Bartlett.* New York: Harry N. Abrams, Inc., 1985.

Storr, Robert. "Wounded Sphinx." *Art in America. 77,* March, 1989.

Strand, Mark, with forward by Robert Hughes. *Art of the Real, Nine American Figurative Painters.* New York: Clarkson N. Potter Inc., Publishers, 1983.

Trapp, Frank Anderson. *Peter Blume.* Rizzoli International Publications, Inc., 1987.

Tuchman, Maurice. *The Spiritual In Art: Abstract Painting 1890–1985.* Los Angeles County Museum of Art. New York: Abbeville Press Publishers, 1986.

Tully, Judd. *Red Grooms and Ruckus Manhattan.* George Braziller, Inc., 1977.

Weinberg, Jonathan. "Paul Cadmus." *Art In America. 76,* March, 1988.

Winner, Viola Hopkins. "Culture and Commentary: An Eighties Perspective." *Art News, 89,* No. 7, September, 1990.

# Paradigm III
# Figure Above Form

▼

You may remember that our first paradigm on expressive approaches to the figure dealt with the artist's need to reconcile interior feelings of beauty with what the artist perceived in the world. The artist began with a largely unarticulated sense of the beautiful in human form, an ideal, and then wrestled with achieving that ideal by way of descriptive norms, or measurements and symmetries that could furnish repeatable clues to the beauty he or she was trying to express.

In the second paradigm we found the artist pulling away from rational, structured methods of accounting for the exterior world and moving instead almost entirely into interior sensibilities. The figure remained part of the artist's expression but mainly as a conjured image, less-than-ideal, within a spatial or emotional illusion, distortion, fragmentation, or discontinuity. Form was whatever the artist might use to sustain the figurative content, all the while going against the conventions of surfaces, perspectives and window-like frames.

*Figure Above Form*, our third paradigm, looks at the icon. In this instance the artist addresses a cognitive concern, meaning "that which comes to be known." Before you can be expressive you need first to define your relationship to what we might call a *charged idea*. Fatherhood, for example. For iconic art, the artist's relationship to the *concept* of fatherhood, goes beyond the personal experience of one's own father. The artist's interest must go to what it is that gives the subject matter, fatherhood, significance not based on experiment or experience, but the *a priori* significance. Experiencing your father as a loving protector or as a power abuser could move you to draw either expression. But for the iconic image you must incorporate the cultural understanding of "fatherhood." The *power of the piece must be greater than the image itself.*

The fatherhood icon is represented by Bill Cosby, George Washington, Robert Young, Marlon Brando's Godfather, Black Elk, King Lear, or the ghost father of Hamlet, to name a few. You can begin building iconic images with signs or symbols that some part of your culture or tradition already honors. What signs or symbols would be associated with each of these fathers?

What you the artist do with those signs or symbols can be brought together in an iconic image inspiring an attitude like veneration or reverence. Most religious icons, like pictures of the saints, prompted veneration. Many cultural icons do something similar: Martin Luther King, Jr. and Mother Jones, for example. But iconic images may stimulate other attitudes besides veneration: playfulness, exaggeration, defamation. And the artist's attitude uses the charged idea—fatherhood, for example—to bring iconic weight to a new charged idea—an incongruity, or unanticipated ingredient—something that defies the sacredness of the first. We call these inspirations *iconoclastic.*

Sante Graziani's *Red, White and Blue Rainbow* uses the symbols of the stars and stripes, a rainbow that approximates the White House dome, and Washington's portrait (originally painted by Gilbert Stuart) painted this time with Ben Day dots, an early printing process. The charged idea, "George Washington as Father," is flattened to a design with eight stars and ten and one half stripes in the segments of the flag, and a simulation of the head of Washington from a dollar bill, a new charged idea. All this is clearly iconoclastic, clearly expressive.

The artist aspiring to make icons acknowledges his or her place within a community and its belief systems. The artist cannot intentionally create an icon, except as a replica, or duplicate, inside a defined visual tradition. Only the shared history of a culture and its religious societies

**Figure III.1** *Sante Graziani.* Red, White and Blue Rainbow. *1970. Acrylic on canvas, 40″ × 40″. Collection of Mr. and Mrs. Lewis A. Davis, Haworth, New Jersey. [Seen in* Art About Art. *Jean Lipman and Richard Marshall. E. P. Dutton, New York, in association with the Whitney Museum of American Art, 1978. pg. 135.]*

defines an icon. But the artist can bring iconic subject matter to bear in being expressive with a *charged idea*.

The two species of the icon we will discuss in this paradigm are the religious icon as it has descended from its Byzantine and largely Christian models, and as it has been expanded within one tribal practice, and the cultural icon, shaped by the activities of a mass media that deals in or conveys *charged ideas*.

The icon could be subdivided further. But our concern is the prompting of expressive uses of iconic material. The issues of this chapter will lead directly to the assumptions behind "conceptual art" dealt with in the next paradigm.

Remember, a true icon embodies a charged idea. The icon is not primarily committed to conveying information. An artist's expression walks a fine line insofar as he or she intends to *make* icons. The finished work is not his or her objective, though the art of the icon must always result in an object. An "aesthetic statement" is not the objective. Rather, the artist deals with a concept, the charged idea, that precipitates some kind of worship—religious devotion or galvanized arousal or stages in between. Responses to icons among believers and nonbelievers moves from icon worshippers, *iconodules*, to icon destroyers, *iconoclasts*. The iconoclast seeks to impact on a fixed attitude. Iconoclastic

works usually do not last as long as the icon itself, though they may significantly change the character of the veneration.

A brief background on the religious icon here might be in order. Icons tended to be images of holy personages of certain societies (monastic communities, for instance). Within Christianity the strongest tradition of icons derived from the Byzantine Churches (Byzantine refers to Eastern Orthodoxy from C.E. 330 to 1450), the Orthodox Churches of Russia and Greece. The painted facial image has some of its roots in Egyptian funerary portraiture. The artists used hot waxes mixed with pigments called encaustic painting. But funerary art was just one of several precursors to the Byzantine icon. The imperial portrait carried in processions was another. And many pre-Christian pagan panels of gods and goddesses survive for historical referencing as well.[1]

The *shapes* of icons were principled in forerunners too. Typically, an icon panel was rectangular or round (known as a tondo panel). Folding diptychs (two panels) had their origins in ancient writing tablets made of wood or ivory. The more popular shape for the icon was the triptych, the dramatic central image covered with two folding wings. Many of these icons were portable, the earliest made for private use.[2]

The *materials* of which icons were made ranged from wood, to ivory, to frescoes, to mosaics, to textiles, to embroidery.[3]

Emperor Leo III mounted a concerted campaign in the middle of the eighth century to destroy "graven images." His son Constantine V set out to destroy every kind of icon but the symbol of the cross. The iconodules were pressed to defend their use. They argued that neither the materials nor the images themselves were worshipped, but "that visible images could show forth invisible truths of religion."[4] People were stimulated by these images to honor the prototypes they represented. Icons were a means of approach, a channel through which one could beseech intercession to whomever was represented. They are still incorporated to similar ends, a tribute to the enduring power of a charged idea. We will return to the Byzantine icon later in this paradigm.

Students should not overlook the precipitating conditions behind icons. For example, few times are more charged than the dying of a great figure. Or the collective release of redemption through a mighty act. Iconic material is never far below the surface of even everyday activity: washing, feeding, embracing, dreaming, birthing.

Not long after icons were restored to a place of honor, c. 843, artists again took up the themes of the "dying savior." One of the most enduring subjects of the Christian faith is the crucifixion of Christ. Matthias Grunewald painted his Crucifixion in the 1500s.

Two of the following examples have images that validate what is already believed. The worshipper stands before Grunewald's huge closed Isenheim Altarpiece acknowl-

*Figure III.2* *Matthias Grunewald (1455–1528).* The Crucifixion. *From the Isenheim Altarpiece (closed). c. 1510–15. Panel, 8'10'' × 10'1''. Musee Unterlinden, Colmar.*

*Figure III.3* *Graham Sutherland.* The Crucifixion. *1946. Oil on hardboard, 96'' × 90''. Church of St. Matthew, Northampton, England. Art Resource, N.Y.*

edging the waning agonizing body of Christ, feeling the pain of the lacerated flesh, believing that his or her own sins were taken away through His death. This image represents only one part of the passion of Christ. The believer prays in faith to the resurrected Christ. Grunewald allows the viewer to worship the Christ beyond the image.

Graham Sutherland's *Crucifixion* is a contemporary icon. Repeating Grunewald's upturned palms, this tormented concentration camp Christ appears to suffer not by way of torn and lacerated flesh, but through the thorny harshness of line and dramatic lights and darks that define and illuminate the body. Yet through this figure a believer still perceives, and prays to, the risen Christ, the iconic expiation of the believer's sin.

On the other hand, Stanley Spencer's *Crucifixion*, while it references iconic material, takes us into the composition behind the sacrificial Christ. We see two men pounding nails into his body, their faces suggesting sadistic amusement. One sinner on another cross is shrieking at Christ. Mary, his mother, like a rag doll, is prostrate on the ground. Observers peer out windows from inner city flats. This crucifixion is not an icon. It is iconographic, but not an icon. We are more involved with human responses to the crucifixion activity than with divine action. Christ's crucifixion is understood, a given. Spencer emphasizes callous human reactions, and he brings to the material a personal narrative. This expressive practice, the narrative, will be discussed in greater detail under the final paradigm, *Figure the Fulfiller of Form.*

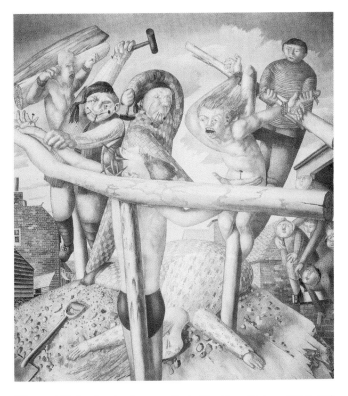

*Figure III.4* *Stanley Spencer, R. A.* The Crucifixion. *1958. Oil on canvas, 216 × 216cm/ 85'' × 85''. Private collection.*

Francis Bacon's *Fragment of a Crucifixion* exploits the intensity of feeling prompted earlier in the Isenheim altarpiece. A dwarfed male figure screaming in terror is slipping into a transparent box space as a fierce pit-bulldog-like apparition scrambles over the top of the cross. Anonymous viewers, as in the Spencer piece, remain in the background. This crucifixion is not an icon either. We are much more aware of the figure's helpless horror incited by an animal, than we are prompted to move through Christ's death. This is iconographic, too, but not an icon.

## A Study:

**Robert Zimmerman's work is a well-conceived contemporary use of the crucifixion icon. Christ is bound in a straightjacket, standing in a flood of tears produced by the student's ancestors who weep, perhaps for Christ's death, for their own sins, for their sense of having disempowered the bondaged Christ, who remains loving nonetheless. The charged idea contained in the dying of a great figure will indeed tend to lead a believer beyond the image.**

Cultural icons typically have some religious overtones. But the charged ideas of a culture most often have to do with the empowering or ennobling memory of a heroic figure or leader, or with the impact of deeply felt archetypes, such as the brave warrior, the wise mother, the innocent child. Artists have also been responsive to sexual models, where the arousing, charged idea is erotic or romantic. Obviously, rendering of the naked figure easily serves such iconic objectives.

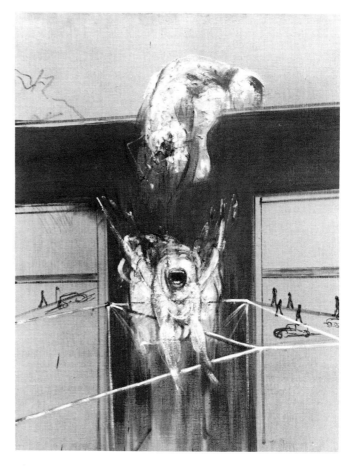

***Figure III.5*** *Francis Bacon.* Fragment of a Crucifixion. *1950. Oil and cotton wool on canvas, 55″ × 42 3/4″. Collection of Stedelijk Van Abbemuseum, Eindhoven, The Netherlands. [plate 31 in* Sacred Art in a Secular Century *by Horton Davies and Hugh Davies. The Liturgical Press, Collegeville, MN. 1978.]*

***Figure III.6*** *Robert Zimmerman.* Crucifixion. *1982. Graphite on rag paper, 9″ × 12″.*

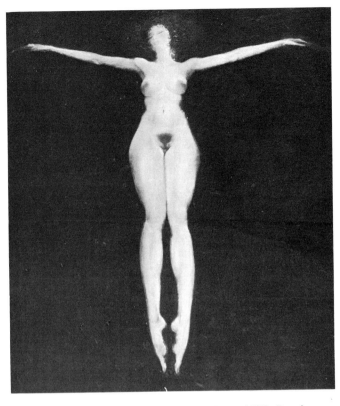

**Figure III.7** *Ernst Fuchs.* Ballerina at Rest. *1970. Pastel on canvas, 39 3/8″ × 39 3/8″, (100 × 100 cm) Ernst Fuchs. Translated by Sophie Wilkins, introduction by Marcel Brion. Harry N. Abrams, Incorporated, Pub., N. Y. 1977.*

**Figure III.8** *Norman Rockwell (1894–1978).* The Peace Corps Led by Kennedy. *Oil Painting for* Look *cover, June 14, 1966. Photo courtesy of the Norman Rockwell Museum at Stockbridge, MA. [Seen in Susan E. Meyer's* Norman Rockwell's People. *Harry N. Abrams, Inc., Publishers, New York, 1981. Page 121.]*

Ernst Fuchs' *Ballerina at Rest* is a cultural cruciform figure. By no stretch of the imagination is this a Christian icon. The title of the work notwithstanding, what we see is a sensuous nude woman lying in a cruciform position. Fuchs presumes sacredness of the naked female form, all the while conferring culturally recognizable sexual overtones through the full virginal breasts and abundant thighs. This particular cultural icon may affirm what a patriarchal society believes about the "woman" it both desires and sacrifices.

John F. Kennedy is among the twentieth century's best-known cultural heroes. Norman Rockwell, a supreme interpreter of American icons, painted *The Peace Corps Led by Kennedy*. The images point to trust and goodness, and to a working-together, idealized populism. They allow one to feel that greater and nobler things can be done if one follows Kennedy's vision, his strength, and tolerance of others.

But cultural iconic material, unprotected by doctrine, is always vulnerable to competition among other charged ideas. John F. Kennedy and his brother, Bobby, were parodied by Lena Svedberg in her 1968 drawing, *The Kennedy Brothers Plus One*, an allusion to Lyndon Baines Johnson, a leader who pursued both men for different reasons. By 1968, when Svedberg's drawing was created, the Kennedy brothers had been tragically assassinated. But political intrigues and personal improprieties surfaced with complications for all three men and for their presumed heroic proportions. In effect, the cruciform figures were caught in Svedberg's drawing with their pants down, hero's halos slipping, LBJ in the background running to climb onto his cross.

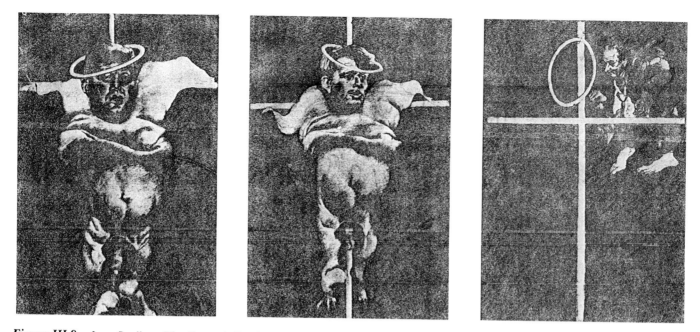

**Figure III.9** *Lena Svedberg.* The Kennedy Brothers Plus One. *3' × 103" × 73". Drawing. Dessin. Zeichnung. 1968 [Arts Council of Great Britain Catalogue]*

The crosses and haloes here are symbols, the traditional iconic symbols of holiness. The way they are used by Svedberg, along with the contorted, abstract figures, aims at political parody. The intent is iconoclastic, with the implicit warning, "Watch out for graven images!"

President Kennedy was assassinated in Dallas, Texas, November 22, 1963, by a man named Lee Harvey Oswald, who himself was shockingly murdered by Jack Ruby as live TV cameras were witnessing. Probably because Oswald was indicted moments before and murdered before millions of viewers in cold blood enroute to his arraignment, citizens rose in a sympathetic outcry against lawlessness and a gun-crazed mindset. The artist, Ed Paschke, used an unidentified newspaper photograph to create his own cultural icon of the accused assassin "protected" by the very symbols in whose name he presumably acted as assassin. Oswald is recognized as a victim as often as he is remembered for victimizing.

The pervasiveness of the photographic, cinematic, and televised image has perpetrated almost instant iconographic identity, and artists are quick to incorporate these images in their work. "Popular culture" contains its own charged ideas. The entertainment hero and heroine are

**Figure III.10** *Ed Paschke.* Purple Ritual. *121.9 × 81.3 cm. (48" × 32") Oil on canvas. 1967. Photo courtesy of the Phyllis Kind Galleries, Chicago and New York. Photo credit William H. Bengtson. Private collection.*

**Figure III.11** *Andy Warhol (American 1930–1987).* Triple Elvis. *1964. Aluminum paint and printer's ink silkscreened on canvas. 82"H × 71"W (208.0 cm × 180.5 cm) Note: Unsigned. Virginia Museum of Fine Arts, Richmond. Gift of Sydney and Frances Lewis. 85.453.*

**Figure III.12** *George Segal.* The Movie Poster. *1967. Plaster, wood, and photograph, 71" × 28" × 36". Collection of Mr. and Mrs. John Powers, Aspen, Colorado. Seen in* George Segal, 12 Human Situations. *Catalogue of an exhibition organized by the Museum of Contemporary Art, Chicago, Ill. May 26, 1988.*

often depicted as cultural icons. Two of the most memorable were (and are) Elvis Presley and Marilyn Monroe. Both of these "stars" captured the yearnings of men and women. Each was viewed as the ideal lover, the partner longed for, hoped for, the god and goddess dream come true in flesh and blood.

Andy Warhol captured the virility, the energy, the commanding man, in his silk screen, *Triple Elvis*. In this work, Elvis is in control, tripled no less, appearing to move. Motion in movies and songs was an Elvis trademark. People still flock to his birthplace museum to listen to his songs, to look and touch and buy memorabilia, to emulate him, even to emulate his somewhat iconoclastic attitude toward social norms.

Marilyn Monroe is remembered in George Segal's famous icon of an icon. A man stands before the movie poster (the icon) of Marilyn, his body language indicating a yearning for the "something" Marilyn could give him; though that "something" is clearly out of reach, it is venerated.

Just as Marilyn Monroe was considered "a man's dream," and Elvis Presley, in his pre-drug years, as "a woman's dream," artist Ron Heinen has drawn two extraordinary icons of the *everywoman* and *everyman* standing before symbolic images of female (sex) and male (beef). The *everywoman* stands next to a poster that affirms she is simultaneously "Rosie the Riveter," taking over as breadwinner when her husband or lover is off to war (WWII)—

**Figure III.13** *Ron Heinen.* Equality, a State of Mind and Media. *1990. Graphite, watercolor, fixative layers on rag paper, 28" × 19 1/2".*

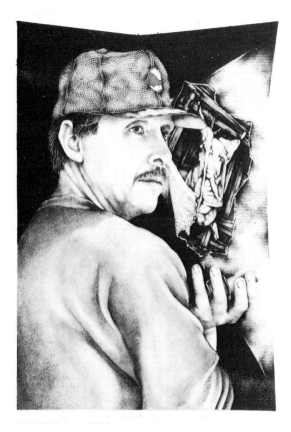

**Figure III.14** *Ron Heinen.* Side of Beef. *1990. Graphite, watercolor, fixative layers on rag paper, 17 3/4" × 12 1/2".*

both worker and provider—and she is a sexual fulfiller, albeit passive and submissive, serving as the classic cultural icon of woman as the nurturing, sexual object.

The male in Heinen's drawing, *Man as Meat*, shows a middle American male, a worker. Next to him on the right is a torn picture of Rembrandt's panel, *The Slaughtered Ox*, 1655, now in the Louvre. In our culture a man is intended to be a "hunk," a "side of beef." The classic icon of man is as the defender, provider object.

### A Study:

**A brief departure here outlining Heinen's procedure might foster notions for your own iconic drawings. Usually Heinen develops his ideas from music lyrics or from remarks people make in conversations. He works with the ideas sometimes for years to find his images. Using photos and geometric sketches to organize line, he draws contour lines with very light cross-contour to develop compositional directions. Painting with the grisaille method, meaning monochrome painting with subjects sometimes having the appearance of marble statues, he covers his drawing. When color is used, it is topical, usually a watercolor or gouache medium. The pointillist technique, a dot-by-dot process, used with a variety of pencils across the whole drawing, achieves an**

**overall uniformity of media, materials, and method. And finally, many layers of spray fix are added to induce a shiny, plastic surface.**

One of our most famous national icons is Grant Wood's *American Gothic*. The pair in the painting are an odd couple. Wood asked his thirty-two-year-old sister, Nan, to pose as the daughter next to the *aging father*, who was Wood's sixty-two-year-old dentist. Together in the painting, they became man and wife, symbols of middle-American farmers empowered with rigid routine, austere simplicity, unchanging habits, backed with uncomplicated religiosity. The cross on the Gothic construction (which was, in fact, a house) behind the couple suggests a country church.[5]

Because this couple is so universally perceived within iconic references, they can speak from many venues—becoming any type of married couple, representing the government, embodying average Americans, becoming proud old folks. All of these charged ideas, however, subject the painting to parody whenever the charged ideas are vulnerable. So often has Grant Wood's piece been treated iconoclastically, it is almost simultaneously an icon and an anti-icon. The portraits have been changed into Nancy and Ronald Reagan, Ku Klux Klanners, a social security poster, an old man and woman with headphones. *American Gothic*

*Figure III.15* Grant Wood (1892–1942). American Gothic. *1930. Oil on beaver board, 76 × 63.3 cm. Friends of American Art Collection, 1930. 934. Photograph © 1991 The Art Institute of Chicago. All Rights Reserved. Art Institute of Chicago.*

*Figure III.16* Buy Some Apples. *From a late 19th-century popular French magazine. [*"Eroticism and Female Imagery in Nineteenth Century Art," *by Linda Nochlin, from* Art News XXXVIII, 1972. Edited by Thomas Hess and Linda Nochlin.]

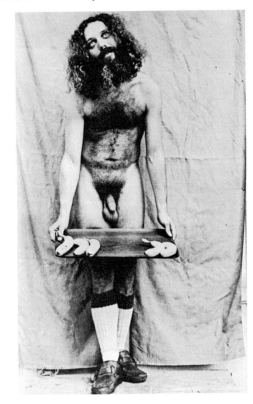

*Figure III.17* Buy Some Bananas. *1972. Photograph.* ["Eroticism and Female Imagery in Nineteenth Century Art," *by Linda Nochlin from* Art News Annual XXXVIII, *1972.]*

seems to have an application as pervasive as the many competing ideas of popular American culture. But even the distortions of this cultural icon, not unlike Duchamp's *Mona Lisa*, outlasts its pejorative uses.

Let's look further into the cultural icon. Linda Nochlin in "Eroticism and Female Imagery in Nineteenth Century Art" illustrated her point about the ways a particular culture can project sexual stereotypes by contrasting two images. One was from a French magazine. A nude woman—apart from hose, shoes and necklace—stands holding a tray of apples. The title is *Buy Some Apples*, a metaphor for her breasts.

Ms. Nochlin's question then was, "What happens when traditional erotic symbols change sex?" Apart from shoes, socks, and a beard, a nude male stood holding a tray of bananas. The title was *Buy Some Bananas*, a metaphor for his penis.

This woman and this man are iconographic symbols, objects that represent something by association, in this case, marketed sex.

Dan Douke, on the other hand, has produced a significant cultural icon with his *Austin Healy Sprite, Etc.* His image allows the male observer the supreme voyeuristic high of woman as machine, sweet, smooth, controllable.

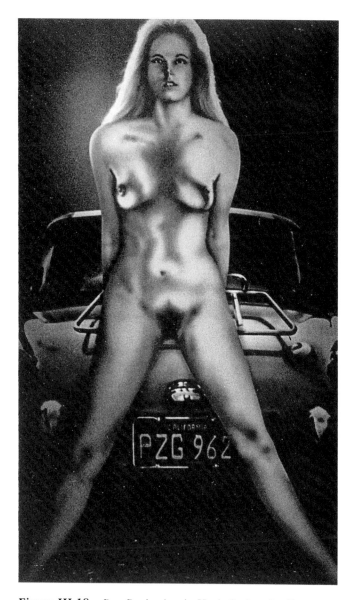

**Figure III.18** *Dan Douke.* Austin Healy Sprite. *Acrylic painting. Nearly life size. Jack Glenn Gallery, Corona del Mar, California. [Seen in* Erotic Art of the Masters, the 18th, 19th, and 20th Centuries *by Bradley Smith. A Gemini-Smith, Inc. Book Published by Lyle Stuart, Inc., 120 Enterprise Avenue, Secaucus, New Jersey 07094.]*

**Figure III.19** *Kay-Lynne Johnson.* The Big 'A'. *1990. Computer-generated image on bond paper, 11" × 8 1/2". [Arnold Schwarzenegger posing in* Flex Muscle, *September 1990, page 76. Photo by Art Zeller. The "Three Flags," 1958, of Jasper Johns is an encaustic on three superimposed canvases, 31" × 45.5". Collection of Mr. and Mrs. Tremaine, N. Y.]*

Business advertising deals richly in iconic ideas. Advertisements similar to Douke's painting are built around the selling of products, in this case, the car. Promoters might assume the object is sold in two ways—to the man who sees cars as erotic and controllable, and to the woman who extends her sexual image by association.

The counterpart to girl and machine is man and muscle. Arnold Schwarzenegger, a titled bodybuilder, shows off his competitive form as a former Mr. Olympia. Schwarzenegger has become synonymous in this country with body building and man as meat. "The body-builder's goal is appearance not action."[6] Women are using men now as men have so long used women. There is, as Margaret

Walters states in her book, *The Nude Male*, "that mix of mockery and unwilling admiration" from male and female audiences, but for different reasons.

An iconoclastic version of Schwarzenegger became inevitable, still with the same pose, but this cultural icon now incorporates the use of the American flag with stripes scrambled and stars as dollar signs.

Coca-Cola is well-established as one of the best known soft drinks in the world. Some people are compulsive consumers of representations of a charged idea. Coke's ad team has been known to show people from many nations holding hands walking in a line up and over a grassy, flowered hill singing in unison, "It's the re-al thing." The inference, of course, is world unification, peace and harmony . . . to the believers who consume the sacred drink. B. Tobey captured a similar manipulative and idealistic concept in his cartoon in the *New Yorker Magazine*.

*Figure III.20* Kay-Lynne Johnson. Payola 'A'. 1990. Computer-generated laser print on bond, 11″ × 8 1/2″. [Photo of A. S. posing in Flex Muscle, *September 1990, page 76. Photo by Art Zeller.]*

Marisol's *Love,* relies on a cultural icon that states much more than its diminished forms appear to say. Your god is whatever you worship.

The religious icon is a shared and sacred image within a particular faith society. One can find religious icons in Buddhism, Hinduism, Christian and other traditions. For example, exceptional iconic material belongs to tribes of Native American Indians.

Some of the earliest Christians were converted pagans who brought their painting heritage to the new task of finding a Christian iconography. Figurative images with specific attributes began being seen in the catacombs. Sacred writings were illustrated, saintly deeds were painted, images bore witness and enshrined tradition and history.

*Figure III.22* Marisol. Love. 1962. Plaster and glass (Coca-Cola bottle) 6¼ × 4⅛ × 8⅛″. Collection, The Museum of Modern Art, New York. Gift of Claire and Tom Wesselmann.

Once Christianity became the official religion of the Roman Empire under Constantine the Great, religious art could become public art.[7]

Subsequently, icons of the Christian faith had pictorial standards set by the Second Council of Nicaea in 787 C.E. About two centuries later, those standards were reasserted in Constantinople, and interpreted by St. John of Damascus, who explained that veneration was not paid to the image, but to the prototype it represented. Remember, the icon served as a means, a channel through which spiritual knowledge and sanctity passed from the image counterpart to the worshipper. Church fathers thought mystical contact needed procedure, rules, and conformation in its imagery.

The iconic image was required to:

1. Be a likeness
2. Be vested
3. Show an inscription
4. Be rendered clearly recognizable
5. Face out of the picture.[8]

The list sets forth the requirements for a figure. There are also icons of deeds, great biblical moments, virtues and avoidable vices, and so on, which expand the subject repertoire of icons. Some of the oldest, most beautiful, and most famous icons in the world are at St. Catherine's Monastery in the Sinai desert. The artists remain largely anonymous, but doubtless developed their craft through the teaching by generations of a distinctive image-making that sought to lead viewers quickly and meaningfully to the charged ideas behind their efforts.

Elijah was revered as a great prophet of the Northern Kingdom in ninth century B.C.E. He shaped the history of his day, and his messages have infused biblical thought since.[9] Our example, painted by one named Stephanos, of our icon includes a raven, harking back to the passage in I Kings 17:2–7, when God sends Elijah into the desert wilderness by the brook Cherith east of the Jordan saying, ". . . you shall drink from the brook, and I have commanded the ravens to feed you there."[10]

The creator of *The Heavenly Ladder of John Climacus* furnished edification of the monks. The "charged idea" is how to get to heaven. At the end of the sixth century, John Climacus was abbot of Sinai. He wrote a treatise on reaching heaven. One *arrived* by climbing 30 rungs of a ladder (probably a reference to Jacob's ladder). The thirty rungs stood for the number of virtues the monks needed to achieve their goal. Temptations caused many of them to fall off, the devils waiting to help. John was the first to make it thus to heaven.[11]

John Climacus lived in a cave some two hours journey by foot from St. Catherine's monastery. The icon was painted five centuries after his death in the first half of the seventh century.

**Figure III.23** *Stephanos.* The Prophet Elijah. *Circa 1200 A.D. 130 × 67 cm (three-quarter life-size) St. Catherine's Monastery, Sinai.*

*Christ* is one of the oldest and best kept icons in the monastery. The icon is located immediately inside the portal doors on the left, for viewing along with many other icons. The *Christ* icon, regarded as bearing a likeness carried truthfully through a carefully preserved oral tradition, depicts a nearly life-size bust of Christ. The frontality of the face, the hand raised in blessing, and the striking immediacy of the portrait is compelling for those who go to St. Catherine's as pilgrims.

**Figure III.24** The Heavenly Ladder of John Climacus. *The second half of the 12th century. 41 × 21.6 cm. The Monastery of St. Catherine's, Sinai.*

**Figure III.25** Christ. *Sixth century A.C.E. Encaustic. 84 × 45.5 cm. The Monastery of St. Catherine's, Sinai.*

**Figure III.26** The Virgin of Vladimir. *Byzantine. Ca. 1131* A.D. *104 × 69 cm. Jahrundert Tretyakov Gallery, Moscow, Russia. Art Resource, N.Y.*

Icons of mother and child are known within many traditions the world over. One of the most famous is the Byzantine (not Russian) icon titled *The Virgin of Vladimir*, the holiest icon in Russia. Looking at the Madonna's eyes, we see that one pupil is to the viewer, the other is glancing away. That device is compelling, not frightening. Faces that stare straight at the viewer can disturb viewers. Portrait photographers are known to ask the sitter to look just above or beside the camera to divert a direct gaze.

The Byzantine standard for heads, as you remember from the first paradigm, incorporated the three-circle scheme, equal measurements from forehead to nose root, to nose base, to chin. In this icon, the mother's head is tilted slightly with her cheek touching the Christ child. The interaction between mother and son shows more sensitivity than many icons of the same subject.

## A Study:

**Susan Henderson used the *Virgin of Vladimir* as a basis for depicting a cultural icon. Placing the Madonna in the same position, she changed the child's figure into a capital "A." Following Byzantine rules, she used an inscription, "Excellence," with the "A" just above it, and repeated the "A" in the border. Her icon was intended as an illustration of the pervasive honoring of the 4.0 grade point average. People seem to worship the "A" itself, not respecting the substantive effort in the achievement. This is iconoclastic, a parody using religious iconographic images to address a cultural idolatry.**

Jacopo Bellini and his two sons were a famous trio of painters in Venice, Italy, in the 1400s. His work, *Madonna and Child with Donor* shows the Madonna turned to her right, the precocious Christ child (very much a baby by comparison to the "little man" child in the Virgin of Vladimir) raising his hand to bless the diminutive donor kneeling in adoration. Bellini's attempt at perspective was in keeping with the times, when the problematical one-point perspective was finally resolved by the architect/artist Brunelleschi in Florence, about 1420. Bellini's image is more a painting of the mother and child than it is an icon. It is iconic, but not an icon in the strictest sense. The kneeling donor, and the perspective and the engagement of the Christ child with the donor, suggest that the expressive intent is not beyond the piece, but strictly within the circumstances of its creation.

## A Study:

**J. D. Larson has abstracted his own iconographic issues from Bellini. Using simple compositional geometric shapes given by mountains and light he has placed Madonna and Child in like positions. The diagonal lines are meant to give a fixed and obvious center of interest, but also to indicate a cross. Larson placed his own hand over his work, tracing around his palm and fingers to suggest "a reaching for belief and the desire to have faith."[12]**

Faith is acted out in many ways. The Navajo concept of the universe requires that all parts—each with a power of good and evil—remain balanced and in harmony. When the balance is upset, physical or mental illness ensues and a curing ceremony is needed. A diagnostician determines the illness and the "chant" (selected from six main group-

**Figure III.27** *Susan Henderson. Excellence Icon. 1985. 15″ × 12″, India ink, gold ink, graphite on rag paper.*

**Figure III.28** *Jacopo Bellini (1400–1470).* Madonna and Child with Donor. *c. 1441. Panel, 23″ × 16″. The Louvre, Paris. Art Resource, N.Y.*

ings) needed to effect the cure. The curing ceremonials, or chantways, are complicated, involve the extended family, last from one to nine nights. Through the rituals the patient is purified and eventually identifies with the deities whose help was sought.[13]

The curative circle is expressed in a sand painting selected from 600 to 1000 designs. The circles range from one to twenty feet in diameter. Once the patient is seated in the middle of the hieroglyphic iconic images, which may have taken a group a full day to complete, the ceremony begins. The disease or evil of the sickness is absorbed by the sand, later ceremonially buried.[14]

Colors usually indicate direction; white/east, yellow/west, black/north (and male) and blue/south (and female). Red is for the sun. The designs can be recreated many times, using a variety of chants.[15]

Many American Indian tribes revere the circle (many other cultures also, e.g., the Buddhist use of the Yin-Yang) and revere kinds of animals. Painting on circular shields was common among Plains Indians. The Arapaho shields often carried a painted turtle. Turtles are hard to kill, and so would the Arapahos be hard to kill whenever they displayed the turtle. These images are clearly iconic, and are often revered in their present form as relics.

Robert Lentz, living in New Mexico, is one of a very few contemporary icon painters. Briefly, Lentz's philosophy is that whenever a cultural group honors a person, an icon of that person is not inappropriate for empowering the believers. Lentz's long apprenticeship at Mt.

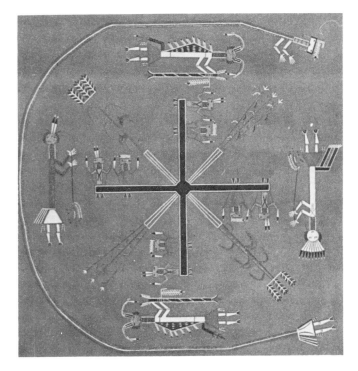

**Figure III.30** Whirling Logs Sandpainting. *This is the theme of the Night and Feather Chants. The four sacred plants of the Navajo are in the center: corn, squash, beans and tobacco. [Seen in Tom Bahti's* Southwest Indian Ceremonies. *Photos by K. C. Den Dooven, 4th printing, 1979, pg. 9. Published by KC Publications, Box 14883, Las Vegas, Nevada 89114. L. C. number 79–136004.]*

**Figure III.29** *J. D. Larson.* Reaching for Bellini. *Spring, 1990. Graphite on rag paper, 15″ × 15″.*

**Figure III.31** *Painted shield. Arapaho. c. 1850. Diameter 19″.* Courtesy of National Museum of the American Indian, *Smithsonian Institution. New York City. Neg. #33645 Buffalo hide shield with deerskin cover; black and green painted decoration on cover with pendant feathers and bells.*

**Figure III.32** *Robert Lentz.* Icon of Kateri Tekakwitha, *1986.* *Gessoed untempered masonite, acrylic paint, 23 karat gold leaf.* *Bridge Building Icons.*

**Figure III.33** *Robert Lentz.* Navaho Madonna. *1988. Gessoed tempered masonite, acrylic paint, 24 karat gold leaf. Bridge Building Icons.*

Athos in Greece disciplined him in his craft. He uses six coats of gesso on tempered masonite. He paints with acrylic paint and 24 k. gold leaf.

Lentz visited the area in Montreal where Kateri Tekakwitha died. "It was a depressed little village on the outskirts of the city. There I found a middle-aged woman who would talk with me. I asked her how she would like to see Kateri painted. The woman replied that no one wanted to see her with a lily for purity. Kateri was an Iroquois her whole life, a Catholic for two years. If anything prepared her for the hardships she underwent in those two years, it was her Iroquois upbringing."[16]

The Iroquois were composed of six nations. Blood feuds nearly destroyed the tribes by mid-fifteenth century. An Indian prophet arose named Dekanawidah. He traveled among the tribes preaching reconciliation and peace. The people agreed, forming a unified government.

The unification symbol was a huge tree under whose branches all peoples could speak peace. An eagle warns the people below whenever peace is threatened. The tree, like the earth, rides on the turtle's back. Kateri holds three objects in one hand, the cross in the other, gifts of the Holy Spirit.

"It's an awkward symbol because the turtle weighs more than Kateri can hold, but things are weightless in icons. This icon means a great deal to the Iroquois people—it's their identity within the icon."[17]

The Navaho people living in northeastern Arizona share their closeness to Christianity in Robert Lentz's *Navaho Madonna.* Beauty is the Navaho ideal and their religion interweaves with experiences of the beautiful. The phrase beneath Lentz's image reads "Beauty has been reestablished." Mary is Navaho, her child strapped to a cradleboard. The inscription above her says "Mother of God," and above the child, "Jesus Christ." The letters in Christ's halo say, "I am who I am." The elongated figure seen on the left, bottom, and right border is a Rainbow Yei, one of the Holy People of the Navaho religion. Rain allows things to grow and it is good.[18]

Let us turn to another kind of young male child, Calvin and his perennial friend, tiger Hobbes. Calvin wants to validate his existence by sharing the event of his sleigh ride with TV viewers. The impact, of course, would be in the image, not the substance. He recognizes with glee that their private adventure could become a cultural icon.

The TV, in and of itself, is more a monolith. Every room could have something that says, "the chair," meaning the throne or "the TV," providing an altar around which there will be rituals.

But the TV becomes an icon when Walter Cronkite is added, that sage man who is remembered by millions for saying, "And that's the way it is . . ."

---

Figure III.34    Bill Watterson. Calvin and Hobbes. © 1990 Universal Press Syndicate. Reprinted with permission. All Rights Reserved.

**Figure III.35**    *Wendy Woloson*. The Monolith. *Lithograph, 9″ × 6″.*

**Figure III.36**    *Wendy Woloson*. Monolith. *Walter Cronkite superimposed.*

**Figure III.37** Close-up view of the Statue of Liberty. *New York [MPC Enterprises Inc., Glendale, N.Y. 11385]*

Another famous cultural icon is the Statue of Liberty given to us by the French. The European vision of America's promise was embracing and receptive, like a female. She stands unblinkingly for liberty, equality and justice for all. She lights the way for the tired, the poor, the refugee. She was the gateway to a better life at the entry point.

### A Study:

**Tim Greenzweig's interpretation of the Statue of Liberty is iconoclastic. The Lady as Liberty means the dollar sign. She emerges from the center of a dollar bill instead of George Washington, the usual image. To have real liberty, in other words, one needs to be wealthy.**

Clowns are a wonderful iconic symbol in our midst, and are indeed, seen all over the world. A clown's face, by the way, is protected by law. Stealing another clown's face is like forging a name on a check.

The "Clown" is a charged idea. He or she is the hero, the anti-hero, the everyman, the everywoman, one of the shadows of each of us. The clown infers humor *and* pathos, both characteristic of our human condition.

Often clowns are nonverbal, miming their way into people's hearts. Clowns can be disruptive, irreverent, free, and playful. Their intent is throwing aside assumptions

**Figure III.38** *Tim Greenzweig.* Liberty. *12″ × 9″. Graphite on rag paper, 1990.*

about ordinary life and practices. The clown in history has been a symbol for the underdog, the lowly, the remnant people. They take incredible risks, like deflecting a charging bull from a fallen rider. They spend great amounts of energy uncovering small things, showing forth hidden treasures, like laughter in people who have forgotten how. Their cherished possessions, their humor, they give to others.[19]

One of our most beloved clowns is Red Skelton, who, apart from his clownish disguises, is a clown at heart. He is a man who seeks the good in people. He is a man who prompts laughter, as he says, "without using four-letter words."

Toulouse Lautrec, a French painter, frequented houses of prostitution, bars, theaters, and sundry places for misfits. Dwarfed because he was dropped as a baby, he felt keenly misplaced all his life. His empathetic work with people gives us iconographic insight into his world of the late 1800s.

His drawings and paintings of the clowness, Cha-U-Kao, show her as she makes her entrance in the Moulin Rouge. Her costume is a ruffled yellow blouse, dark green bloomer pants, and stockings. She wears white makeup and a tall white wig tied with dangling yellow ribbons. Evidently she had the ability to disrupt any situation at the flick of a finger, making things revolve around her, yet she hardly seemed to be working.[20]

**Red Skelton**

E. P. Dutton · New York

***Figure III.39*** Red Skelton. *E. P. Dutton, New York.*

***Figure III.40*** *Toulouse Lautrec (1864–1901). Clown. 1896. Lauros-Giraudon. Art Resource, N.Y.*

## A Study:

Two student illustrations of clowns move to iconoclasm in that they both represent the clown "laughing on the outside and crying on the inside." The first example shows a rodeo clown pretending to laugh or enjoy what he is doing all the while trying to repel an advancing bull. One face in the background indicates the clown's fear, the other face intends to show the normal person behind the mask.

Not unlike Emmett Kelley, one of our great clowns, our next example is a hobo clown, down and out, and nurturing the bottle of booze highlighted in his right hand. The contradiction again is toward the iconoclastic, the unmasking of the benevolent icon of the clown.

Remember, in creating icons you work with a "charged idea." You work with a concept that precipitates worship—religious devotion—and galvanizes arousal, or elicits stages in between. The power of the icon is arrested by the image, not teased by it. The personal expressive statement here is not what counts as much as the artist's conceptual statement. The esthetics of the piece lead the viewer past the piece. True icons encapsulate the most charged notions of a culture.

We will close this chapter with one of the greatest cultural icons in the world, the Mona Lisa, made an icon not

**Figure III.41** *Marcia Moore.* Rodeo Clown. *9′ × 12′. Colored pencil on rag, tracing paper overlays.*

**Figure III.42** *Robert Zimmerman.* Hobo Clown. *9″ × 12″. Graphite on rag paper.*

204

*Figure III.43*   *Leonardo da Vinci (1452–1519). Mona Lisa. c. 1503–05. Panel, 30 1/4″ × 21″. The Louvre, Paris, France. Scala/Art Resource, N.Y.*

*Figure III.44*   *Marcel Duchamp (1887–1968). L.H.O.O.Q. [Seen in* The World of Marcel Duchamp *by Calvin Tomkins. Time-Life Editions, New York, 1966. Pg. 61. Mary Sisler Collection, New York.]*

by the artist, but by the centuries of reverence for the way the portrait embodies "high art," the very best the Renaissance could produce, the height of beauty and mystery. Leonardo never expected the lady with the mysterious smile (computer artist Lillian Schwartz believes Mona's face is Leonardo's)[21] to permeate lives the way she has, but European and Western cultures have given her the adoration associated with "icon." The ethos of the Mona Lisa's charged idea held for five hundred years, until Marcel Duchamp prompted another kind of perception by drawing a graffiti-like mustache on her. Duchamp made his point about the tendency to exaggerate aspirations, and his parody served a new charged idea without diminishing the original. Duchamp was a gifted iconoclast who did not take anything away from the Mona Lisa, all the while creating a climate for reconsidering the concept of "masterpiece."

## ENDNOTES

1.  Kurt Weitzman. 1978. *The Icon, Holy Images—Sixth to Fourteenth Century.* New York: George Braziller, pp. 7–14.

2.  Ibid.

3.  Ibid.

4.  Harold Osborne. Ed. 1970. *The Oxford Companion to Art.* Oxford: Clarendon Press, p. 554.

5.  Wanda M. Corn. 1938. *Grant Wood the Regionalist Vision.* New Haven and London: Yale University Press, Published for the Minneapolis Institute of Arts, pp. 136–142.

6.  Margaret Walters. 1978. *The Nude Male, a New Perspective.* New York and London: Paddington Press, Ltd., p. 294.

7.  P. Francastel. 1967. *Medieval Painting, Vol. II.* Edited by Hans L. C. Jarre. Translated by Robert Erich Wolf. Laurel Edition, pp. 7–11.

8. Harold Osborne. *The Oxford Companion to Art.* Oxford, England: The Clarendon Press, pp. 553–555.

9. Madeleine S. Miller and J. Lane. 1973. *Harper's Bible Dictionary.* New York: Harper and Row, p. 157.

10. The Holy Bible, Revised Standard Version. 1962. Philadelphia, Penn.: A. J. Holman Company, I Kings 17: 3–4, p. 342.

11. John Galey. *Sinai and the Monastery of St. Catherine.* Introduction by George H. Forsyth and Kurt Weitzmann. Massada, 1979, p. 93.

12. An interview with J. D. Larson, Summer, 1990, Iowa State University, Ames, Iowa.

13. Tom Bahti. 1979. *Southwestern Indian Ceremonials.* Las Vegas, Nevada: KC Publications, pp. 6–10.

14. Ibid, p. 10.

15. Ibid, p. 10.

16. An interview with Robert Lentz, September 29, 1990.

17. Ibid.

18. From an information card by Robert Lentz, Bridge Building Icons, P.O. Box 1048, Burlington, VT. 05402, 1988.

19. From *A Clown is Born* brochure. Mass Media Ministry, Baltimore, Maryland.

20. Horst Keller. 1968. *Toulouse-Lautrec: Painter of Paris.* Translated from German by Erika Bizzarri. New York: Harry N. Abrams, Inc., p. 70.

21. Cynthia Goodman. 1987. *Digital Visions, Computers and Art.* New York: Harry N. Abrams, Inc., p. 82.

## BIBLIOGRAPHY

Bahti, Tom. *Southwestern Indian Ceremonials.* Las Vegas, Nev.: KC Publications, 1979.

Benezra, Neal. *Ed Paschke.* New York: Hudson Hills Press, 1990.

Brooks, Rosetta. *Art Forum.* "Leon Golub Undercover Agent." January, 1990, Volume 28, 5, pp. 114–121.

Campbell, Joseph. *Myths to Live By.* Foreword by Joshua E. Fairchild, New York: Viking Press, 1972.

Carline, Richard. *Stanley Spencer, R.A.* Royal Academy of Arts, Over Wallop, Hampshire, England: BAS Printers Limited, 1980.

Casson, Hugh; Carline, Richard; Causey, Andrew. *Stanley Spencer, R.A.* Royal Academy of Arts, London: Weidenfeld and Nicholson, 1980.

Christe, Yves; Velmans, Tania; Losowka, Hanna; and Recht, Roland. *Art of the Christian World, A Handbook of Styles and Forms A.D. 200–1500.* New York: Rizzoli, 1982.

Cleland, Charles. *The Art of the Great Lake Indians.* Flint Institute of Arts, 1973.

Compton, Michael and Copplestone, Trewin. *Pop Art (Movements of Modern Art.)* London: Hamlyn Publishing Group Limited, 1970.

Corn, Wanda M. *Grant Wood, The Regionalist Vision.* New Haven: Published for The Minneapolis Institute of Arts by Yale University Press, 1983.

Davies, Horton, and Davies, Hugh. *Sacred Art in a Secular Century.* Collegeville, Minnesota: The Liturgical Press, 1978.

Feder, Norman. *American Indian Art.* New York: Harry N. Abrams, Inc., Publishers, 1971.

———. *Ernest Fuchs.* Introduction by Marcel Brion, translation by Sophie Wilkins, New York: Harry N. Abrams, Inc., 1979.

Francastel, P. Edited by Hans L. C. Jaffe, with translation by Robert E. Wolf. *Medieval Painting, Volume II.* New York: Dell Publishing Company, Inc., with Harry N. Abrams, Inc., 1967.

Galey, John. *Sinai and the Monastery of Saint Catherine.* Introduction by George H. Forsyth and Kurt Weitzmann. Massada, 1979.

Garrard, Rose and Nairne, Sandy. *Women's Images of Men Catalogue.* Institute of Contemporary Arts, London, 1980.

Gilmour, Pat. *The Mechanized Image, An Historical Perspective on 20th Century Prints.* Arts Council of Great Britain. Twickenham, England: C.T.D. Limited, 1978.

Graves, Robert. *The Greek Myths: 1.* London and New York: Penguin Books Limited and Viking Penguin Inc., 1960.

Graves, Robert. *The Greek Myths: 2.* London and New York: Penguin Books Limited and Viking Penguin Inc., 1960.

Hagen, Charles. *Artnews.* "All That Jazz." February 6, 1990, Volume 89, No. 2.

Hayes, John. *The Art of Graham Sutherland.* Oxford: Phaidon Press Limited, 1980.

Hess, Thomas B. and Ashbery, John, editors. *Artnews Annual XXXVIII.* "Pinup and Icon." New York: Newsweek, Inc., 1972.

Hochfield, Sylvia. *Artnews.* "Up From the Underground." May 9, 1990, volume 89, No. 5.

*Holy Bible,* Revised Standard Version. Philadelphia, Penn: A. J. Holman Company, 1962.

Inverarity, Robert Bruce. *Art of the Northwest Coast Indians.* Berkeley: University of California Press, 1950.

Janson, H. W., with Janson, Dora Jane. *History of Art.* Englewood Cliffs, N.J.: Prentice-Hall, Inc., and New York: Harry N. Abrams, Inc., 1964.

Keller, Horst. *Toulouse-Lautrec: Painter of Paris.* New York: Harry N. Abrams, Inc., Publishers, 1968.

Mathews, Tom. *Newsweek.* "Fine Art or Foul?" July 2, 1990.

Meyer, Susan E. *Norman Rockwell's People.* New York: Harry N. Abrams, Inc., 1981.

Miller, Madeleine S. and J. Lane. *Harper's Bible Dictionary*. New York: Harper and Row, 1973.

New York Graphic Society Limited. *The New Yorker, Album of Art and Artists*. Greenwich, Conn.: The New Yorker Magazine Incorporated, 1970.

Nochlin, Linda. "Eroticism and Female Imagery in Nineteenth Century Art." *Artnews Annual XXXVIII*, Edited by Thomas Hess and Linda Nochlin, New York: Newsweek, Inc., 1972.

Osborne, Harold. *The Oxford Companion to Art*. Oxford, England: The Clarendon Press, 1970.

Panofsky, Erwin. *Studies in Iconology; Humanistic Themes In the Art of the Renaissance*. New York: Harper and Row, 1962.

Pierre, Jose; translated by W. J. Strachan. *Pop Art, An Illustrated Dictionary*. Great Britain: Published by Eyre Methuen Limited, 1977.

Smith, Bradley. *Erotic Art of the Masters, the 18th, 19th and 20th Centuries*. Secaucus, N.J.: Bemini-Smith, Inc., Syle Stuart, Inc.

Tomkins, Calvin. *The World of Marcel Duchamp*. New York: Time-Life Editions, 1966.

Wagstaff, Sheena, Exhibition Curator. *Comic Icono-Clasm*. Exhibition Organized by The Institute of Contemporary Arts. Sponsored by Mont Blanc. London: 1987–1988.

Walters, Margaret. *The Nude Male: A New Perspective*. London and New York: Paddington Press, Limited, 1978.

Watterson, Bill. "Calvin and Hobbs." *Sunday Des Moines Register Newspaper*, Sunday, February 4, 1990.

Weitzmann, Kurt; Chatzidakis, Manolis and Radojcic, Svetozar. *Icons*. New York: Alpine Fine Arts Collection Limited, 1980.

Weitzmann, Kurt. *The Icon: Holy Images-Sixth to Fourteenth Century*. New York: George Braziller, Inc., 1978.

Wilkins, Sophie, translation; Brion, Marcel, introduction. *Ernst Fuchs*. New York: Harry N. Abrams, Inc., 1979.

Paradigm IV

# Figure and Form as Paradox

▼

**P**aradox here stands for that strain of art that questions what art is and what it is not. If a coin showed any **conventional** art on one side, any **unconventional** art on the flip side, then paradox is seeing both sides at once.

Somewhere between ancient objective canons and current subjective outlooks, another territory of art derives from the turning of conventional beliefs upside down. That contrary artistic territory is *conceptual* (in the mind) as much as *perceptual* (in the eye), which ultimately means you do not need to have a picture to have a work of art.

Using paradox, an artist challenges the viewer's preconceptions about what is actually seen with the eye, turning him or her away from representations to discover the sense in nonsense and the non-sense in sense. Art can be said to reside as much in the act of discovering as in the art object itself.

When do figure and form function as a paradox? A visual paradox is a meeting of opposites. Paradox will show contradictory information, yet identify a truth. One or the other, figure or form will lead to expectations the other appears to contradict.

Paradox raises dilemmas that force the suspension of logical expectations to wait on another meaning. For instance, in the following example we see Escher's "Hands" drawing each other. How can a drawn hand draw itself?

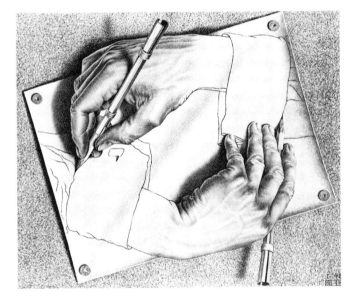

***Figure IV.1*** *M. C. Escher (1898–1973).* Drawing Hands. *1948. Lithograph, 11 1/4″ × 13 3/8″. National Gallery of Art, Washington, Cornelius Van S. Roosevelt Collection.*

## A Study:

**Brian Heydn uses paradox in his drawing by transforming the negative value shape around the outline of his elevated hands into a bird in flight. The "correlative other," his hands, lose their shape as the bird becomes more pronounced.**

Philosophers have approached the experiencing of dilemma by way of the illustration of two boxes, the sentences within each contradicting the other.

**Figure IV.2** *Brian Heydn.* Hands to Flight. *1990. Graphite, 18" × 4".*

| | |
|---|---|
| The sentence in Box B is true. | The sentence in Box A is false. |
| Box A | Box B |

**Figure IV.3** *A Paradox.*

Similarly, a story is told of Epimenides (from the Isle of Crete), living about the sixth century B.C., who is credited with saying, "All Cretans are liars." If what he said was true, his statement is also false. If correct, then wrong.

This is a "Catch-22," a double bind. Hardly a person is living who has not been caught in a paradoxical situation. But paradoxes raise mystery. They aggravate. They precipitate. They *foster change.*

## A Study:

**Sometimes paradox simply stops a function, seen in our next two examples. Paul Guy renders the fork and spoon useless by bending and mashing the metal items. They cannot serve as eating utensils. They are not now what they seem to be.**

## A Study:

**The board game in our next example cannot be played. With one die, the player runs into a square, sooner or later, which cannot be occupied nor jumped over.**

## A Study:

**Linda Baechler redesigned a shoe with tall women in mind.**

**Figure IV.4** *Paul Guy.* The Paradox of Fork and Spoon.

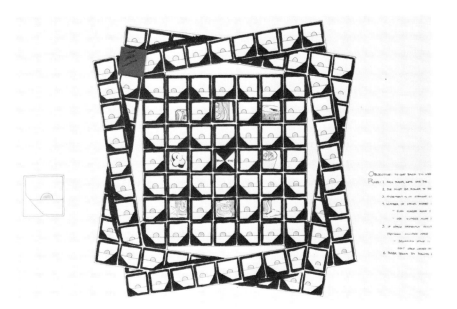

**Figure IV.5**   A Paradoxical No-Win Game.

**Figure IV.6**   *Linda Baechler and Tim Greenzweig.* A High-Heeled Shoe for Tall Women. *Graphite, 18″ × 24″.*

## A Study:

**The next illustration becomes a narrative paradox of a "Family Portrait." Although Jan Schwichlengberg, who drew the back side of a picture, did not include this description, one could add to the paradoxical nature of the image by using the following narrative: In the middle of the picture are my grandparents; to the right are my mom and dad; below and to the left are my cousins from my dad's side of the family, Tommy, Jerry and Buster, the one stringing gum from his mouth. My big brother and I are on the right side. My big brother, Tennessee, my other brother, Texas, my younger brother, Vermont, and me, Montana. My folks gave us the names of their favorite states. Of course, my big brother goes by "Tom," my other brother goes by "Tex," my youngest brother likes to be called "Monty," and I go by "Teeny Anna" since I'm the teeniest.**

In the early 1950s Robert Rauschenberg approached Willem deKooning to ask a favor. Rauschenberg asked deKooning for an important drawing that he, Robert, could erase. The result is shown in the example.

"What possessed the man?" one might ask. Rauschenberg asserted an intellectual dilemma through visual means. The drawing was a deKooning. Rauschenberg erased the crayon on paper over a period of one month. Only a few stains are left, imbedded in the paper. The final piece is, indeed, a Rauschenberg, the strokes of the erasure as medium removed deKooning's marks—almost, but not quite. "I was trying . . . both to purge myself of my teaching and at the same time exercise . . . possibilities as I was doing monochrome no image."[1]

This is an act of creating art by removing art. We are asked paradoxically to imagine a lost image while at the same time to appreciate a new one through the action of losing it. Not unlike life. Think of yourself for a moment. You are not now what you were at the age of two. That *child* is no longer. It has been constantly, daily, replaced by something new. Your physical, mental, emotional, and spiritual erasers have been *forming* you, the *figure* since you were born. Paradoxically, you are more you in the act of losing yourself.

Let's turn, for just a moment, to some other fields of inquiry. Paradox itself has been with us, of course, since the race began to dwell on incongruities. When paradox elicits tension, it suspends belief. That characteristic allows paradox to be dramatic in nature. Theater is one of the better known arenas for paradox. You may sit in a theater

**Figure IV.7** *Jan Schwichlenberg.* Family Portrait. *1982. Charcoal, graphite, and chalk on grey paper, 12″ × 10 1/2″.*

**Figure IV.8** *Robert Rauschenberg.* Erased deKooning. *1953. Erased pencil, 19″ × 14 1/2″ (48.3 × 36.9 cm). Collection of the artist/Leo Castelli Gallery.*

for two hours on a Saturday evening in the year 1995. The play you watch, while actually lasting two clock hours, might, between Act I and Act III cover someone's lifetime. As you are affected by the drama of the play, you can feel suspended from your own time frame, transported into the time frame within the play, and realize a profound truth from that play that can affect your beliefs. Is your present-time reality illusion?

The Buddhist system of thought contains the paradox of Nirvana, the word for the release from the cycles of life.[2] Nirvana in a positive conception is like the further shore, the refuge; it is peace. On the other hand, in a negative sense, it is the Void. It is Cessation. But Buddhism denies there is an individual soul. The human being disintegrates at death. Who or what is left to attain Nirvana?

Swiss psychologist Carl Jung uses paradox to comment on the origins of Christian faith. "Oddly enough the paradox is one of our most valued spiritual possessions, while uniformity of meaning is a sign of weakness. A religion becomes impoverished when it cuts down its paradoxes; but their multiplication enriches because only the paradox comes anywhere near to comprehending the fullness of life. Non-ambiguity and non-contradiction are one-sided and thus unsuited to express the incomprehensible." Jung goes on, "As witness to this we have Tertullian's avowal: 'And the Son of God is dead, which is worthy of belief because it is absurd. And when buried He rose again, which is certain because it is impossible.'"[3]

In mathematics, Cantor's proof that there is no greatest number ensures that there is no smallest fraction. There is no beginning. Something is always a fraction of something else.

The famous "*Cogito ergo sum*," "I think, therefore I am," denoted the end of a process by which Rene Descartes (1596–1650) set out to question the truth of everything by rational means. He found he could doubt all truths except the fact that he was doubting.

Before we look at works of art and suggested "*Studies*" by which you may illustrate paradox and its off-shoots, we might acknowledge that the term *paradox* is derived from two Greek words, *para*, meaning "beyond," and *doxa*, meaning "opinion." Paradox flirts with all of the terms listed below, some more, some less.

Paradox—a meeting of opposites, which can become a vehicle for truth.

Enigma—an obscure riddle, one that is puzzling, inexplicable.

Ambiguity—a condition susceptible to multiple interpretations. The condition is equivocal, meaning it is deliberately unclear or misleading.

Obscurity—suggests meaning hidden in difficult form.

Vagueness—signifies a lack of definite form.

Cryptic—suggests a puzzling terseness intended to discourage understanding.

Irony—the use of words or images to convey the opposite of their literal meaning. ("Beautiful weather!" said when it's raining.)

Contradiction—an assertion or expression of the opposite.

Remember the ambiguity of E. G. Boring's work showing the beautiful young girl or the old woman? The image you see depends on your perception of the configurations of lines and shapes.

Be aware that the person "who habitually plays with paradoxes is too often suspected of doing intellectual parlor magic for the lay mind. The truth more often is that he is hot after realities."[4]

Of the twentieth century masters, Marcel Duchamp, the chaser of paradox, unassumingly transformed our expectations for Western art. Remember his Mona Lisa with a mustache and a beard? Its title was *L.H.O.O.Q.* Read phonetically in French this title means "she has a hot ass." That reproduction on which Duchamp drew helped reshape art dramatically. He trashed a highly respectable convention for the purpose of "de-estheticizing" artistic convention.

Duchamp was *the* seminal influence for *mobiles* (movable sculpture) and *readymades* or found objects. He was *a* seminal influence in Dada, Surrealism, Junk Sculpture, Assemblage, Box Art, Color Field Paintings, Minimal Art, Pop

***Figure IV.9*** *Roger Cheney, from* Catalogue Illusions. *1974. Arts Council of London. Devised by E. G. Boring.*

**Figure IV.10** *Marcel Duchamp (1887–1968). L.H.O.O.Q. 1919. Rectified Readymade: pencil on a reproduction, 7 3/4″ × 4 7/8″, Private Collection, Paris. [pg. 129, Anne D'Harnoncourt catalogue,* Marcel Duchamp, *MoMA/Philadelphia Museum of Art, Distributed by New York Graphic Society Ltd., Greenwich, Connecticut, 1973.]*

**Figure IV.11** *Christopher Adams. 1991 drawing (graphite, 20 × 15″) based on Marcel Duchamp's "With my tongue in my cheek," 1959. The original was a mixed media work of cast plaster depicting a portion of the actual jaw, mouth, and cheek of Duchamp which was superimposed over his own pencil drawing on paper screwed onto a wooden panel, 25 × 15 × 5.1 cm. The original is in the collection of Robert Lebel, Paris, France.*

Art, Op Art, and Conceptual Art. Duchamp delighted in paradox, in puns, in pitting verbal against visual, in double and triple meanings.

Duchamp lived a life of paradox. As he said to Harriet and Sidney Janis, "I have forced myself to contradict myself in order to avoid conforming to my own taste."[5] He wanted not so much anti-art, but "anaesthetic," something without judgement. He wanted his art to be para-doxa, beyond opinion. For him, "the game (Art) becomes an intellectual liberation. It upholds no taste; it is a conceptual domain operating according to its own rules . . . derived from a fascination with the conflict rather than with its resolution . . . to see the sense in non-sense and the nonsense in sense . . . this of course means that the spectator must become an artist. The spectator has to play the game with Duchamp."[6]

Duchamp constantly tries to keep the viewer from taking both him and his art seriously. Seriousness, to Duchamp, was a flaw in conventional art. The puns in his titles were a road map of ambiguity.

Duchamp sometimes used another name, a woman's. "Rrose Selavey" comes from two other words, "arrose" and "c'est la vie'." This alter ego is almost as famous as Duchamp's *Mona* in *L.H.O.O.Q.*

Duchamp rebelled against the "retinal" response to art. "All through the last half of the 19th century in France there was an expression, 'as stupid as a painter'," Duchamp said. "And it was true—that kind of painter who just puts down what he sees *is* stupid . . ."[7]

In his move away from retinal art, Duchamp was influenced by the concept of motion, a radical departure from the static positions of most figures in art. The cinema was barely in its second decade. Marey's chronophotographs and Muybridge's photos of motion studies, were important influences, along with Cubism, leading to Duchamp's most famous painting, *Nude Descending a Staircase.*

With *Nude Descending the Staircase,* Duchamp had taken a major step away from retinal art toward conceptual art.

*Figure IV.12* Duchamp as Rrose Selavy. *Photographed by Man Ray, c. 1920–1921 [pg. 17, Anne D'Harnoncourt catalogue, Marcel Duchamp in New York, c. 1920–21.]*

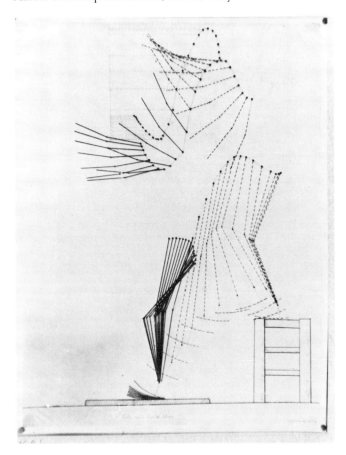

*Figure IV.13* *Jules Etienne Marey (1830–1904).* Jump From a Height With Stiffened Legs. *Musee Makney.*

*Figure IV.14* *Marcel Duchamp (1887–1968).* Nude Descending a Staircase, No. 2. *1912. Oil on canvas, 57 1/2″ × 35 1/16″ (146 × 89 cm). Philadelphia Museum of Art, Louise and Walter Arensberg Collection.*

The year 1912 was a turning point for conceptual art, via Duchamp. He quietly helped initiate a revolution by asking what art was. Basically Marcel Duchamp was trying to give *Form* to the concept and not to the object, *Figure*. Giving form to the concept meant emphasizing influences, associations, a train of thought, where the issue was not one of beauty or esthetic permanence but one of process.

He and a group of his friends began developing the radical and ironic ideas which led to a movement called "Dada." In his book, *Dada, Surrealism and Their Heritage,* William Rubin says, "At the heart of Dada lay the 'gratuitous act,' the paradoxical, spontaneous gesture aimed at revealing the inconsistency or inanity of conventional beliefs."[8]

Duchamp helped upend common assumptions about art. One of the "upends" was his notion of the *Readymade,* a term he defined in 1915 some two years after he had fastened the now famous bicycle wheel on a kitchen stool.

Much later, in 1961, Duchamp gave a speech, "Apropos of Readymades." He referenced several early, important ideas—the Readymades, Readymades-aided, Reciprocal Readymades, limited production of Readymades because of indiscriminate repetition, works of assemblage, the prerogative of choice, and anesthetic taste.

### Apropos of Readymades

*In 1913 I had the happy idea to fasten a bicycle wheel to a kitchen stool and watch it turn.*

*A few months later I bought a cheap reproduction of a winter evening landscape, which I called "Pharmacy" after adding two small dots, one red and one yellow, in the horizon.*

*In New York in 1915 I bought at a hardware store a snow shovel on which I wrote 'In Advance of the Broken Arm.'*

*It was around that time that the word "Readymade" came to my mind to designate this form of manifestation.*

*A point which I want very much to establish is that the choice of these "Readymades" was never dictated by an esthetics delectation.*

*This choice was based on a reaction of* visual *indifference, with at the same time, a total absence of good or bad taste . . . in fact a complete anesthesia.*

*One important characteristic was the short sentence which I occasionally inscribed on the "Readymade."*

*That sentence instead of describing the object like a title, was meant to carry the mind of the spectator towards other regions more verbal.*

*Sometimes I would add a graphic detail of presentation which in order to satisfy my craving for alliterations, would be called "Readymade Aided."*

*At another time, wanting to expose the basic antinomy between art and Readymades I imagined a "Reciprocal Readymade," and used a Rembrandt as an ironing board.*

*I realized very soon the danger of repeating indiscriminately this form of expression and decided to limit the production of "Readymades" to a small number yearly. I was aware at that time, that for the spectator even more than for the artist, art is a habit forming drug and I wanted to protect my "readymades" against such a contamination.*

*Another aspect of the "Readymade" is its lack of uniqueness . . . the replica of a "Readymade" delivering the same message; in fact nearly every one of the "Readymades" existing today is not an original, in the conventional sense.*

*A final remark to this egomaniac's discourse:*

*Since the tubes of paint used by the artist are manufactured and ready made products we must conclude that all the paintings in the world are "Readymades aided" and also works of assemblage.*

Marcel Duchamp 1961[9]

A little earlier in this paradigm you may remember the statement that thought for Duchamp, was the *form* and the object was the *figure* or about the figure. Art became *the comment on* the human figure and not the image of the figure itself. At the 1917 exhibition of the New York Society of Independent Artists, Duchamp submitted a porcelain urinal titled *Fountain,* and on its back signed "R. Mutt." The group did not take him seriously. Duchamp made a social gesture—he resigned from the Society—to heighten the seriousness of the non-serious. If art is choice—and the word art in its origins means 'to make,' activating choices—Duchamp chose a urinal, chose to turn it upside down, chose to call it *Fountain,* was rejected from the show and chose to resign from the art association. He chose to move art outside of art.

What has this to do with drawing? Not much just yet. We are simply exploring conceptual art, meaning art that has more to do with thinking processes than images. Conceptual art *did* open doors to new imaging as we will soon see.

This way of thinking about art might be confusing at first, but if you take the time to digest what was happening through one artist's work, perhaps you can begin to participate in it. Even though the process might not make sense for you, try to stay with the process, most especially if it is new to you. Eventually you will be able to think *conceptually.* And, hopefully, those kinds of insights will lead you to more sophisticated art.

Duchamp moved from the art object to the nonobject, *air,* whose existence was determined by the ampoule (a sealed glass vial) surrounding it. Duchamp declared its contents. "This is the precious ampoule of fifty cubic centimeters of AIR OF PARIS, I brought back to the Arensbergs in 1919."[10]

***Figure IV.15*** *Marcel Duchamp (1887–1968). Air de Paris. (50 cc of Paris Air), 1919. Ex coll: Louise and Walter Arensberg, New York, gift of the artist in 1920. Philadelphia Museum of Art, The Louise and Walter Arensberg Collection.*

Let's clarify this idea a little farther. The *Readymade* meant taking an object out of its context and putting it into a new space that created a new thought and a new meaning entirely for that object. That specific object could never again be what it seemed to be.

With the ampoule of air, Duchamp had provided a container for air in the form of a cultural artifact, *someone's* glass-blown vial. There is no figure here, although *figure is inferred*. The vial is the object. The object serves to heighten the *concept, the form*, which enables the viewer to engage with the idea of the "magical air of Paris," without even seeing or breathing it. Once seen or breathed, it is gone.

Further, each of us as human beings can be understood to be a *primary form* as a container for air. We have lungs and breathe. So, each of us becomes a Readymade at birth. In a strange way, we have revisited *figure as form*.

Again, what Duchamp did with Readymades was *to infer*, to reference a human and figurative presence. Duchamp had created a *concept about the figure*, no longer requiring the figure itself. This climate allowed the figure to be expressed in infinite variety both in "effect" (an outcome or result) and "affect" (a movement of emotions). What the *object says about the figure* becomes important now.

For example, a hangman's noose or an electric chair suggest a powerful figurative presence, either in anticipation of a human context, or as a reminder.

Turning now to the off-shoots of paradox listed earlier, enigma and mystery, for example, other artists surface, generally with images as end results. Some artists realized that even the essential form could be hidden.

The 1960–70s artist Christo built on the ideas of Man Ray, an artist of the 1920s. Christo simply wrapped things. The wrapped form lends mystery to familiar space and its contents. The viewer plays with enigmatic meanings. *Package on Wheelbarrow*, remains benign in title, but reminds one of a struggling, wrapped body, setting up psychological tensions within a viewer.

A spin-off from both Readymades and Cubism was the collage. Collages are made up of found objects too. The collage can also be created from found images, an approach Kurt Schwitters used in *For Kate*. Duchamp would have called these *Readymades Aided*.

***Figure IV.16*** *Christo.* Package on Wheelbarrow. *1963. Cloth, rope, wood, and metal, 35 inches high × 60 inches long × 23 inches wide. Collection the artist, New York. [Seen in William S. Rubin's* Dada, Surrealism and Their Heritage, *MoMA, New York, 1968].*

***Figure IV.17*** *Kurt Schwitters.* For Kate. *1947. Collage of pasted papers 4 1/8″ × 5 1/8″. Collection Mrs. Kate T. Steinitz, Los Angeles. [pg. 57 in* Dada, Surrealism, and Their Heritage *by William S. Rubin. MoMA, New York, 1968.]*

## A Study:

Collage allows the incorporation of many extraneous items into a work. Selecting the pieces from what you have found or torn is challenging because each "piece of space" has its own dimensions (lines), textures, and colors. Try to find "stuff" that works together in a drawing. Using thick mounds of colored rubber cement usually looks and smells bad. Achieving unity within your piece is important.

For instance, in our example, the "stuff" this student selected was heavy drawing paper, watercolor paper, watercolors, pastels, and grease crayons. She used frottage, sometimes called "rubbing," by laying a paper over a textured surface and running the pastel stick over the paper to leave the textured imprint. Her imprint appeared like a rough weave.

The rough-weave texture prompted her to think of weaving actual torn strips of paper. From an old watercolor and an old drawing she cut and tore both wide and narrow strips.

Once she visually redescribed the model lying on the table with those pieces of paper by drawing and rearranging the torn pieces of frottaged paper, she began to weave actual strips of paper for the surrounding space. The piece is unified in materials, media, image, and mode. Instinct, common sense, and creativity can move hand in hand.

Remember that much of what the Dadaists did was simply to "displace," to take something out of context and place it in another. The visual Surrealists wanted to push beyond that. They elaborated on existing objects, often pushing them to dream images or *meta-physical* images. Looking at Magritte's *The Eternal Evidence*, with its discrete frames we can conjecture that he was interested in parts and wholes. The question might be, Is the sum of the parts greater than or equal to the whole?

**Figure IV.19** *Rene Magritte (1898–1967).* The Eternal Evidence. *1930. Painting: oil on five separately stretched and framed canvases—each canvas from top A 22.2 × 12.4 cm; B 19.4 × 24 cm; C 27.1 × 19.2 cm; D 22.2 × 15.9 cm; E 21.9 × 12 cm. Courtesy of the Menil Collection, Houston, Texas.*

**Figure IV.18** *Netha Pike.* Lady on a Bed of Frottage. *1989. 18″ × 24″. Collage, frottage, pastels, and oil-base crayons.*

Another aspect of enigma emerged when artists moved into performance art, which in the 60s was called a Happening. A "Happening" was a theater event, usually without script. Even when there was a play, it had no beginning, middle, or end. Performances for artists liberated them from the art object.[11] Audience participation seemed important to the performance artist. The Happening broke the stationary form of "the picture on the wall." The roots of Happenings go back to the Dadaists. Duchamp helped his Surrealist friends put two major art exhibitions together, one in Paris, one in New York. For both unsuspecting viewing audiences, these were "Happenings."

The first occasion was the 1938 International Surrealist Exhibition in Paris. After the artists had created works of art (static objects) Duchamp hung them for display. Within the gallery a mock furnace with an electric lamp inside kept the ceiling lit but the art works in near darkness. Gallery-goers (the audience) were given flashlights and a new experience. They used their flashlights to pick their way through the heavy layer of leaves on the floor and the huge beds in the corner to see the art works.

In the second installation in 1942 in New York, the layout for the exhibition, "First Papers of Surrealism," *hindrance* instead of *promoting* was at work again. Duchamp used miles of string to tie up the entire room preventing people from getting close to the works.

**Figure IV.20**    Gallery-Goers With Flashlights. *International Surrealist Exhibition, Paris, 1938. [pg 161 in* The World of Marcel Duchamp, 1887–1968 *by Calvin Tomkins and the Editors of Time-Life Books, 1966.]*

**Figure IV.21**    Sixteen Miles of String. *October, 1942. Layout for the exhibition, "First Papers of Surrealism." Organized by Andre Breton and "his twine" Marcel Duchamp on the premises of the Coordinating Council of French Relief Societies, New York. [pg 23. d'Harnoncourt's* catalogue, Marcel Duchamp, Phila. Mus. of Art.]

**Figure IV.22** *Max Ernst (1891–1976). Loplop Introduces. 1932. Pasted papers, watercolor, pencil frottage, and photograph, 19 5/8″ × 25 3/8″. Collection Mr. and Mrs. E. A. Bergman, Chicago, Illinois.*

What kind of sense is that? Well, if you attend an international art show, and are prevented from viewing the art, what is left is seeing each other. That paradoxical situation prompted people to turn to one another and communicate. And what greater work of art than the human being, our figure, with all our differences?

Turning now to one of the more studio-oriented, image-making off-shoots of paradox, "Chance," we'll give a brief introduction.

We're back to Duchamp again, who started us off in this concept by dropping three strings. He held three threads one at a time horizontally over individual painted canvas strips, which in turn were mounted on glass panels. One at a time he dropped the threads and glued them where they dropped.

"This experiment was made in 1913 to imprison and preserve forms obtained through chance, through my chance . . . because my chance is different than your chance . . . ."[12]

There are many types and kinds of Chance. Each one of the following you might like to try:

*Frottage*: Along with incorporating "found objects," which retained their own identity but were used as part of a spatial arrangement in a collage, Max Ernst used a technique called *frottage*, or "rubbing."

## A Study:

Place a piece of paper over a number of textured surfaces, each time running a pencil, charcoal, conte crayon, or whatever over the paper lying on the textured surface. If enough rubbed shapes overlap, you might discern an image. The resulting image rests largely with the laws of chance. Once those shapes begin to form something, you can reorganize and foster new and unforseen images into a drawing that might appear unorthodox. But the image will decidedly carry your special "stamp." No one else's chance and final decisions are like yours.

*Decalcomania* was first exploited in 1936. Yves Tanguy's "Decalcomania," is one such experiment.

## A Study:

Spread paint, gouache, oil, acrylic, or watercolor on a sheet of paper. Place another sheet of paper on top and press down arbitrarily in different places. Peel off the top sheet. Marks appear that can suggest many kinds of images such as landscapes, mountains, rocks, people, and so on. See what images you can find, if any, showing through the unplanned markings. Remember to turn your work to each of the four sides of the paper so that you can gauge which side yields the most meaningful and workable image. If there is

something there, work back into the piece with your medium to finalize the image, drawing in more lines or shapes. Be careful that you do not get "heavy-handed," using so much control that you lose what was good and spontaneous early on. The work below was done with diluted printers' inks. Once an image was seen, drawing strokes helped shape the face and shoulder of the figure.

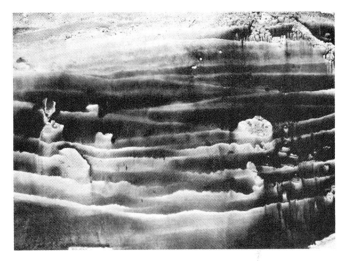

**Figure IV.24** *Yves Tanguy (1900–1955).* Decalcomania. *1936. Gouache, 12″ × 19″. Collection Marcel Jean, Paris [pg. 139,* Dada, Surrealism, and Their Heritage, *William S. Rubin. MoMA, 1968]*

**Figure IV.23** *Student drawing using frottage.*

**Figure IV.25** *A decalcomania drawing.*

*Figure IV.26*  *Fumage done with smoke and wax from a candle.*

*Figure IV.27*  *A student coulage.*

Wolfgang Paalen invented *fumage*, drawing with the smoke left from a burning candle.[13]

### A Study:

**You will need to experiment with several things—types and kinds of candles and papers as well as the angles for holding the paper over the flame. Using just the candle flame smoke is a bit tricky. The picture literally can go up in smoke. Keep working at the process. The "tricks" are not a big problem. You'll learn to control the process, but probably not the image that might ensue.**

Gordon Onslow Ford invented *coulage* (pouring), not to be confused with collage.[14] Pouring allowed pools, with their own edges and formings within a seeming formlessness. Our student example below again used diluted printers' inks. The pools ran together in such a way that a head was formed on top of a background already drawn.

Jackson Pollock, one of the great "pourers," poured paint "from a can, from sticks, hardened brushes, and from large basting syringes."[15] And, Pollock positioned himself on all four sides of a format.

In Pollock's figurative work one confronts the impact of lines while with equal impression the lines remain paint. Obviously the images are not a 'contoured' line as seen in representational work, but imposed into space and onto space simultaneously. "The viewer must confront the physical coextension of paint and image."[16]

This throws art back to the viewer. It is his or her chance too. Questions of figure and form are the viewer's responsibility. Does art happen without a viewer? For Duchamp and Pollock, at least, it did not.

Pollock worked straight from the unconscious, and his comment is worth repeating, " . . . I don't care for 'abstract expressionism,'" Pollock replied once, when asked how to label his painting. "It's certainly not 'nonobjective,' and not 'nonrepresentational' either. I'm very representational some of the time, and a little all of the time. But when you're painting out of your unconscious, figures are bound to emerge."[17]

Probably the single most productive process for students in classes using "Chance" is done with smooth paper, litho crayon or oil base crayons, and a hard-edged plastic eraser.

**Figure IV.28** *Jackson Pollock (1912–1956). Brown and Silver I. c. 1951. Enamel and silver paint on unprimed canvas, 57″ × 42 1/2″. Thyssen-Bornemisza Collection, Lugano, Switzerland.*

**Figure IV.29** *Dana Kunze.* A Chance Drawing.

**Figure IV.30** *Shelli Schubert.* A Chance Drawing.

The following six illustrations were done by students in Life Drawing classes using "Chance."

## A Study:

Use artist's crayons, including litho crayons. Use paper with a smooth tooth (surface) like oaktag (which allows the greasy crayon to streak over the surface when you stroke with the eraser). Begin by drawing three separate images from a model who takes different poses. Draw each image one on top of the other, after rotating the paper ninety degrees for each consecutive pose.

Once the images are drawn, use a 1″ × 2″ plastic eraser held so that one sharp edge can be stroked through the greasy crayon on the paper, smearing the drawn images. Do not *erase* the image. Use the eraser to stroke through the crayon marks, "pulling" the original marks into other kinds of lines and shapes. Like Pollock, rotate the work as you apply some pressure to stroke over and over, pulling the drawn crayon marks.

**Figure IV.31**   *Adam Blyth.* A Chance Drawing.

**Figure IV.32**   *Dave Mottet.* A Chance Drawing.

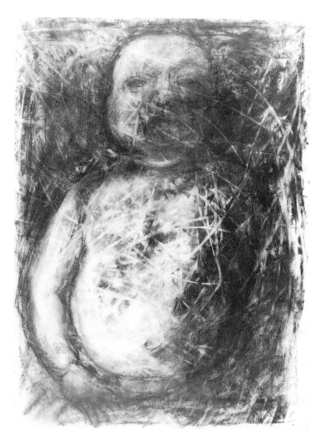

**Figure IV.33**   *Deanne Heying.* A Chance Drawing.

*Figure IV.34* *Peter Hansen.* A Chance Drawing.

When the original image is no longer recognizable, put the drawing on a wall, step back, look at it, and try to see if another image is visible. Rotate the drawing once again, studying each rotated format looking for an image either in close, middle, or far distance. Sometimes there are no other images. Often, happily, there are.

Once one is found, "Chance" is finished and conscious decisions are made about how to "pull" the new image further into existence. The new image is "pulled" into a better definition by identifying the special shapes or figures you want to use. Apply more value, color, or line. Results can be amazing.

All of the examples you see exist because a student drew from a nude model in a classroom, smeared the superimposed images, and captured a much more profound figurative image in the process.

Earlier in this century another spin off from "Chance" was verbal babble, autonomic nonsense language, not unlike the visual doodle. Non-sense language, verbal automatism, was used sometimes in pictures, sometimes in environments. Because 'chance language' is outside the purview of this book, I would suggest for those who are interested to read further in *Dada, Surrealism, and Their Heritage* by William S. Rubin.

The student of figure and form is cautioned, however, by Breton, definer of Surrealism, about what is meant by *chance:* " . . . subjective emotion, however intense, is not directly creative in art; it has value only in as much as it is returned and incorporated indirectly in the emotional source from which the artist must draw."[18]

In other words, a baby's belch is expressive, although it says little except that gas was on the tummy. Which means, of course, that just placing colors on the page in an emotional state does not necessarily a work of art make. Finalizing a work needs many choices.

Another form of "Chance" emerges from Duchamp's art-by-proxy. Duchamp had, in 1919, sent his brother-in-law a geometry book with instructions to hang it by strings from his apartment balcony so the rain and wind could, over a period of time, disintegrate it.

Proxy art can be traced to many Renaissance studios and, no doubt, to Athenian sculptors. But the total removal of the signator's presence and final choice was novel with Duchamp. Pop artist Andy Warhol continued the tradition. Op artist Victor Vasarely laid out his concepts while his apprentices followed his instructions to create the work.

Recently David McKee paid $26,000 for "the right to execute an idea by the artist Sol LeWitt." The formula, in part, with the intermediary using a black pencil, went as follows: . . . "Ten thousand lines about ten inches long, covering the wall evenly." The example on Christie's wall at the auction house in New York, was done by a man who, with pencil and ruler dutifully fulfilled the demands. The day after the auction, the wall piece was destroyed. However, the "original"—the formula—was preserved and belongs to McKee."[19]

And, another spin-off from Chance is the earlier mentioned use of three dropped strings. Duchamp called that piece *Three Standard Stoppages.* The dropped threads whose chance arrangement on canvases were glued onto canvas strips, were also glued to glass strips. Tracing the line created by each thread, wooden slats were cut to shape a hard edged version of each string dropped. Finally the whole assembly was placed together in a box, and both *Assemblage* and *Box Art* were given birth.

**Figure IV.35** *Marcel Duchamp (1887–1968).* Boite-en-Valise (Box in a Valise). *1941–42. [Reproduced courtesy Philadelphia Museum of Art, Louise and Walter Arensberg Collection. Photo by A. J. Wyatt. 16 1/8″ × 14 3/4″ × 4 1/8″ in closed valise.]*

"Here again, a new form of expression was involved. Instead of painting something new, my aim was to reproduce the paintings and the objects I liked and collect them in a space as small as possible. I did not know how to go about it. I first thought of a book, but I did not like the idea. Then it occurred to me that it could be a box in which all my works would be collected and mounted like a small museum, a portable museum, so to speak."[20]

A box is part of our everyday spatial language—cradles, boxcars, dice, jack-in-the-box toys, houses, chests, coffins, to name a few. Not until the twentieth century has the box-as-art taken hold as a genre. Again, the figure is experienced only by referencing.

What the box (the *form*) does is call attention to the special tradition of "framing" the piece of art, giving it a heightened, though often decorative outside contour. The box further plays on the notion of perspective because its sides furnish vanishing points that change with the viewer's angle. Finally, the box can be carried (*the figure*).

John Fraser's work, "Postes Vaticane II," is part of a series of prayer boxes reminiscent of portable reliquaries. The woman, romanticized as an ideal female, may be viewed as a symbol of all men's desires for mother/lover. This boxed assemblage is meant to function as a talisman. The title, "Poste Vaticane II," relates directly to an Italian stamp the artist had since a boy. He referenced the title with the woman, ensuring that "pure love" was based on a belief system, that is, the Catholic faith. The form implies that the work folds in half. It does not actually fold, but the piece infers that one could make an envelope of love and carry it anywhere.

Trova is famous for his *Falling Man Series*, this time through a series of grids into infinity, within a mirrored box.

But Trova in his *Folding Man* of 1968, turns the humanized robotic hermaphroditic figure into foldable units like the box in which it is contained.

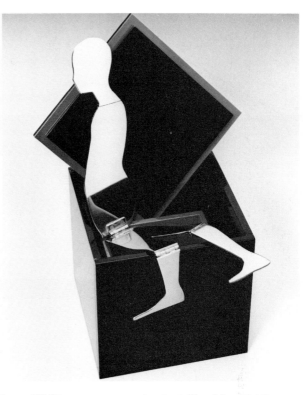

*Figure IV.36*    *John Fraser. Postes Vaticane II. 1986. A prayer-box assemblage.*

*Figure IV.38*    *Ernest Trova. Study/Falling Man (Folding Man). 1969. Solid brass hinged figure in plexiglass box. 5 × 5 × 5″ plexiglass box with 12″ brass hinged figure. Courtesy of the Trova Foundation, St. Louis, Mo. Stamped and numbered in an edition of 100. Published by Multiples, Inc., N.Y.*

*Figure IV.37*    *Ernest Trova. Study/Falling Man (Diminishing Figure) 1964. 16 1/8″ × 14 3/8″. 17 3/4″ Silkscreen on glass, formica photo lithography. Edition 8. Courtesy of the Trova Foundation, St. Louis, Missouri.*

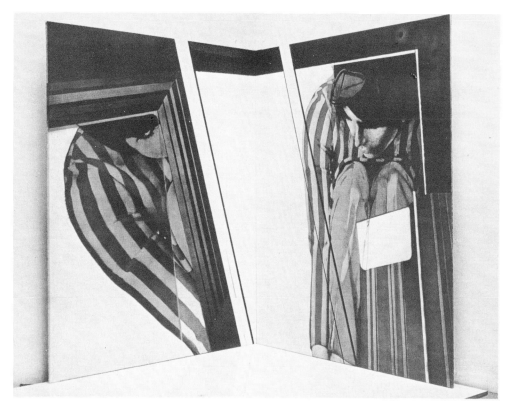

***Figure IV.39*** *Pol Mara.* Out of the Corner of One's Eye. *1968. Oil on canvas and aluminum montage. 2 panels 195 × 162 cm each. Private collection, Los Angeles.*

And Pol Mara's *Out of the Corner of One's Eye* boxes the viewer in by placing two paintings into a corner of a room.

In this brief survey of paradox and its variations in art, we will only touch on Duchamp's use of the *Infra-mince*, sometimes translated *infra-thin*. The principle of infra-mince in action is a *minimum of activity* teeming with creative invention and ideas. An infra-thin can be . . . "When the tobacco smoke smells also of the mouth which exhales it, the odors/marry by infra thin . . . Painting on glass/seen from the unpainted side/gives an infra/thin . . . pastel of dandruff/fallen from the hair/onto a paper/wet with glue . . . cast shadow/oblique. . . . The warmth of a seat (which has just/been left) . . ."[21] The *figure* is alluded to, suggested, inferred. The *form* is a shaped but associated sensation.

In the Morris example below, the figure is alluded to, the form is the sculpture and its associated sensations.

Anne d'Harnoncourt says this about Duchamp's *Infra-mince* in her introductory comments to the catalogue for Marcel Duchamp's Museum of Modern Art Show, later seen at the Philadelphia Museum of Art.

"Duchamp practiced a unique form of aesthetic economy. . . . Consistently turning his attention to the slightest or least-regarded of phenomena, he developed the elusive category that he called 'Infra-mince' . . . by compiling examples: the faint sound made by velvet trouser legs brushing together, the difference between the space occupied by a clean, pressed shirt and the same shirt, dirty.

***Figure IV.40*** *Robert Morris (American, 1931– ).* Table and chair for Sydney Lewis. *1973. Copper table: 91.4 × 122. × 83.8 cm (36" × 48" × 33"); chair: 91.4 × 47.6 × 40.6 cm (36" × 18 3/4" × 16"). Unsigned, Virginia Museum of Fine Arts, Richmond. Gift of Sydney and Frances Lewis, 85.425.1/2.*

*Figure IV.41* Tom Wesselmann (American, 1931– ). Great American Nude #35. 1962. Oil, polymer, and mixed media on board, 48″ × 60″ (121.9 × 152.3 cm). Signed on reverse, upper left: Gan. #35/4′ × 5′ Wesselmann '62. Virginia Museum of Fine Arts, Richmond. Gift of Sydney and Frances Lewis, 85.454.

Infra-mince is explored specifically in a few isolated notes, including the proposal for a 'transformer intended to utilize little wasted energies' (like laughter, the fall of tears, the exhalation of tobacco smoke)."[22]

And Duchamp's influence can be seen in minimal art, which almost always had a strong geometric foundation, a linear square with one corner not quite closed; for instance, a Duchampian concern to "put painting once again at the service of the mind."[23]

And finally we come to Pop Art, that period of art whose artists celebrated banal, everyday objects and events. Wesselman stressed a patriotic red, white, and blue. He used found objects, the bottles and the windows "retrieved from a gutter."[24] He juxtaposed a cropped reproduction of the famous Mona Lisa with a flattened, stylized reclining nude.

Duchamp could never have guessed what the iconoclastic *L.H.O.O.Q.* unleashed. The paradox of the woman with the mustache and goatee is that she is not what she seems to be, and vice versa.

## ENDNOTES

1. Robert Rauschenberg. National Collection of Fine Arts, Smithsonian Institution, Washington, D.C., 1976. p. 75.

2. Robert Lawson Slater, 1951. *Paradox and Nirvana*, A Study of Religious Ultimates with Special Reference to Burmese Buddhism, Chicago: University of Chicago Press, pp. 3–4.

*Figure IV.42* Marcel Duchamp (1887–1968). L.H.O.O.Q. 1919. Rectified Readymade: pencil on a reproduction, 7 3/4″ × 4 7/8″, Private Collection, Paris.

3. Violet Straub de Laszlo, Ed., 1959. *The Basic Writings of C.G. Jung*, New York: The Modern Library, p. 446.

4. Howard A. Slaate, 1968. *The Pertinence of the Paradox (The Dialectic of Reason-in-Existence)*, New York: Humanities Press, Inc., p. 94.

5. Anne d'Harnoncourt and Kynaston McShine, Eds., 1973. *Marcel Duchamp*, Greenwich, Connecticut: The Museum of Modern Art and Philadelphia Museum of Art, distributed by the New York Graphic Society, Ltd., p. 35.

6. Ibid, p. 132.

7. Calvin Tomkins and the editors of Time-Life Books, 1966. *The World of Marcel Duchamp, 1887–1968*, New York: Time-Life Books, Time, Inc., p. 9.

8. William S. Rubin, 1968. *Dada, Surrealism, and Their Heritage*, New York: The Museum of Modern Art, Publishers. Distributed by New York Graphic Society, Ltd., Greenwich, Conn., p. 12.

9. Pierre Cabanne, 1971. *Dialogues with Marcel Duchamp*, translated from the French by Ron Padgett, New York: The Viking Press, plate 10.

10. d'Harnoncourt, *Marcel Duchamp*, p. 291.

11. H. H. Arnason, *History of Modern Art*, Third Ed. Englewood Cliffs, N.J.: Prentice-Hall, Inc., p. 566.

12. d'Harnoncourt, *Marcel Duchamp*, p. 273.

13. Rubin, *Dada, Surrealism, and Their Heritage*, p. 139.

14. Ibid, p. 140.

15. Elizabeth Frank, 1983. *Jackson Pollock*, Modern Masters Series. New York: Abbeville Press, Cross River Press, Ltd., p. 89.

16. Ibid, p. 93.

17. Ibid, p. 95.

18. Rene Passeron, 1978. *Phaidon Encyclopedia of Surrealism*, Oxford: Phaidon Press Limited, p. 47.

19. Michael Small, "Talk About Lines!," *Arts Magazine*, p. 44.

20. Arturo Schwarz, *The Complete Works of Marcel Duchamp*, New York: Harry N. Abrams Inc., p. 513.

21. Paul Matisse, 1983. *Marcel Duchamp, Notes*, Boston: G. K. Hall and Company, selected quotes from notes: 1–46.

22. d'Harnoncourt, *Marcel Duchamp*, p. 37.

23. Ibid, p. 35.

24. Frederick R. Brandt, 1985. *Late 20th-Century Art*, Richmond, Vir.: Virginia Museum of Fine Arts. Seattle and London: Distributed by the University of Washington Press, p. 200.

# BIBLIOGRAPHY

Ades, Dawn. *Dada and Surrealism Reviewed*. Introduction by David Sylvester and a supplementary essay by Elizabeth Cowling. Arts Council of Great Britain, 1978.

Arnason, H. H. *History of Modern Art*. 3rd ed. revised and updated by Daniel Wheeler. Englewood Cliffs, N.J.: Prentice Hall, Inc.

Ausband, Stephen C. *Myth and Meaning, Myth and Order*. Macon, Georgia: Mercer University Press, 1983.

Barr, Alfred H., Jr., ed. *Fantastic Art, Dada, Surrealism*. New York: Museum of Modern Art by Arno Press, December, 1968.

Bool, F. H., Kist, J. R., Locher, J. L., and Wierda, F., *M. C. Escher: His Life and Complete Graphic Work*. Edited by J. L. Locher. New York: Harry N. Abrams, Inc., 1982.

Brandt, Frederick R. *Late 20th-Century Art*. Virginia Museum of Fine Arts, Seattle: Distributed by the University of Washington Press, 1985.

Cabanne, Pierre. *Dialogues with Marcel Duchamp*. Translated from the French by Ron Padgett. New York: The Viking Press, 1971.

Campbell, Joseph. *The Hero with a Thousand Faces*. Bollingen Series, XVII, Pantheon Books. New York: Bollingen Foundation, Inc., 1949.

Cargile, James. *Paradoxes*. A Study in Form and Predication, Cambridge: Cambridge University Press, 1979.

Cavendish, Richard, ed. *Mythology (an illustrated encyclopedia)*. New York: Rizzoli International Publications, Inc., 1980.

Davis, Sandra L. *Predicting Outcome of Paradoxical and Self-Control Interventions from Resistance and Freedom of the Target Behavior Among Procrastinators*. Dissertation, Iowa State University, Ames, Iowa, 1984.

d'Harnoncourt, Anne, and Kynaston McShine, ed. *Marcel Duchamp*. The Museum of Modern Art and Philadelphia Museum of Art, Greenwich, Conn.: Distributed by the New York Graphic Society, Ltd., 1973.

Duchamp, Alexina S., and Paul Matisse. *Marcel Duchamp, Notes*. Arrangement and Translation by Paul Matisse. Preface by Anne d'Harnoncourt. Boston, Mass.: G. K. Hall and Company, 1983.

Escher, M. C., and J. L. Locher. *The Infinite World of M. C. Escher*. New York: Abradale Press/Harry N. Abrams, Inc., 1984.

Forge, Andrew. *Rauschenberg*. New York: Harry N. Abrams, Inc., 1972.

Franck, Sebastian. *280 Paradoxes or Wondrous Sayings*. Translated and introduced by E. J. Furcha. Volume 26: *Text and Studies in Religion*. New York: The Edwin Mellen Press, 1986.

Frank, Elizabeth. *Jackson Pollock*. Vol. 3, Modern Masters Series. New York: Abbeville Press, 1983.

Hallie, Philip P. *The Paradox of Cruelty*. Middletown, Conn.: Wesleyan University Press, 1969.

Herbert, R. T. *Paradox and Identity in Theology*. Ithaca, New York: Cornell University Press, 1979.

Hogarth, William. *Analysis of Beauty*. Edited by Joseph Burke. Oxford: Clarendon Press, 1955.

Hosmer, Rachel, and Alan Jones. *Living in the Spirit.* San Francisco, Calif.: Harper and Row, 1979.

Hulten, K. G. Pontus. *The Machine as Seen at the End of the Mechanical Age.* The Museum of Modern Art. New York Graphic Society Ltd., Greenwich Connecticut, 1968.

Hutcheon, Linda. *A Theory of Parody, the Teachings of Twentieth-Century Art Forms.* New York: Metheun, Inc., 1985.

Jaffe, Michael. *Rubens and Italy.* Ithaca, New York: Cornell University Press, 1977.

Jung, C. G. and C. Kerenyi. *Essays on a Science of Mythology.* Translated by R. F. C. Hull. The Bollingen Library, New York: Pantheon Books, 1949.

Kelsey, Morton T. *Myth, History and Faith, the Remythologizing of Christianity.* New York: Paulist Press, 1974.

Kennedy, Gerald. *The Lion and the Lamb, Paradoxes of the Christian Faith.* New York: Abingdon-Cokesbury Press, 1950.

Kierkegaard, S. K. *A Kierkegaard Anthology.* Edited by Robert Bretall. New York: Modern Library, 1946.

Lucie-Smith, Edward. *American Art Now.* Oxford: Phaidon Press Limited, 1985.

Martin, Robert L. ed. *The Paradox of the Liar.* New Haven, Conn.: Yale University Press, 1970.

McGlashan, Alan. *Gravity and Levity.* Boston, Mass.: Houghton Mifflin Company, 1976.

Mogelon, Alex, and Norman Laliberte. *Art in Boxes.* New York: Van Nostrand Reinhold Company, 1974.

Muybridge, Eadweard. *The Human Figure in Motion.* New York: Dover Publications, 1955.

Napier, A. David. *Masks, Transformation, and Paradox.* Berkeley: University of California Press, 1986.

New York Graphic Society, Ltd. *Arts Since Mid-Century (The New Internationalism, Volume 2: Figurative Art.* Foreword by Jean Leymarie. Greenwich, Conn., 1971.

Niebuhr, H. Richard. *Christ and Culture.* Harper and Brothers Publishers, 1951.

Nietzsche, Friedrich. *The Birth of Tragedy and the Genealogy of Morals.* Translated by Francis Golffing. New York: Doubleday, 1956.

Otto, Rudolph. *The Idea of the Holy: An Inquiry into the Non-Rational Factor in the Idea of the Divine and Its Relation to the Rational.* Translated by John W. Harvey. London: Oxford University Press, 1952.

Parola, Rene. *Optical Art, Theory and Practice.* New York: Reinhold Book Corp., 1969.

Passeron, Rene. *Phaidon Encyclopedia of Surrealism.* Translation by John Griffiths. Oxford: Phaidon Press Limited, 1978.

Pierre, Jose. *Pop Art, An Illustrated Dictionary.* Translated by W. J. Strachan. Chevalier Des Arts Et Lettres. London: Eyre Methuen, Ltd., 1977.

Pincus-Witten, Robert. *Postminimalism.* New York: Out of London Press, Inc., 1977.

Pirandello, Luigi. *Three Plays.* Translated by Robert Rietty, Noel Cregeen, John Linstrum, Julian Mitchell, The Master Playwrights. "Six Characters in Search of an Author." London: Meuthuen, London, Ltd., 1979.

———. *Robert Rauschenberg.* National Collection of Fine Arts, Smithsonian Institution, Washington, D.C., 1977.

Rose, Bernice. *Jackson Pollock, Works on Paper.* The Museum of Modern Art. New York Graphic Society, Greenwich, Conn., 1969.

Roukes, Nicholas. *Plastics for Kinetic Art.* New York: Watson-Guptill Publications, 1974.

Rubin, William S. *Dada, Surrealism, and Their Heritage.* The Museum of Modern Art. The New York Graphic Society, Ltd., Greenwich, Conn., 1968.

Ruhrberg, Karl. *Twentieth Century Art.* New York: Rizzoli International Publications, Inc., 1986.

Schwarz, Arturo. *Marcel Duchamp.* New York: Harry N. Abrams, Inc., 1975.

Schwarz, Arturo. *The Complete Works of Marcel Duchamp.* New York: Harry N. Abrams, Inc.

Siegel, Jeanne. *Artwords, Discourse on the 60's and 70's.* Ann Arbor, Mich.: UMI Research Press, 1985.

Slaatte, Howard A. *The Pertinence of the Paradox (The Dialectics of Reason-in-Existence).* New York: Humanities Press, Inc., 1968.

Slater, Robert Lawson. *Paradox and Nirvana, a Study of Religious Ultimates and Special Reference to Burmese Buddhism.* Chicago: The University of Chicago Press, 1951.

Soby, James Thrall. *Rene Magritte.* Museum of Modern Art, New York: Doubleday, 1965.

Staub de Laszlo, Violet, ed. (with an introduction). *The Basic Writings of C. G. Jung.* New York: The Modern Library, 1959.

Stoppard, Tom. "Artist Descending a Staircase." *Four Plays for Radio,* London: Faber and Faber, 1974.

Stoppard, Tom. *Jumpers.* New York: Grove Press, Inc., 1972.

Tomkins, Calvin, and the editors of Time-Life Books. *The World of Marcel Duchamp, 1887–1968.* New York: Time, Inc., 1966.

Torczyner, Harry. *Magritte, Ideas and Images.* Translated by Richard Miller. New York: Harry N. Abrams, Inc., 1977.

Weisinger, Herbert. *Tragedy and the Paradox of the Fortunate Fall.* East Lansing: Michigan State College Press, 1953.

Wilson, Simon. *Pop.* New York: Barron's Woodbury, 1978.

Wisdom, John. *Paradox and Discovery.* Oxford: Basil Blackwell, 1965.

Wolgast, Elizabeth Hankins. *Paradoxes of Knowledge.* Ithaca, New York, and London: Cornell University Press, 1977.

# Figure: The Fulfiller of Form

▼

This expressive paradigm concerns the telling of tales, the correlation of *image* and *story*. Images that accompany written words tend to fulfill "a priori" forms, meaning that some literary forms—biography, history, fiction, etc.—often precede the visual form. Literary narrative forms usually have artistic merit in themselves, and can shape our imaginations and expectations.

Building fine art upon literary forms has been loudly discredited at times, with attitudes not unlike those some musicians have for the scoring of music for motion pictures. How original, some wonder, is a collaborative art?

Why not tell a story in pictures? Much of fine art has been busily preoccupied with itself over recent decades, looking to its own concepts, trying to remain strictly visual. But if art is sufficient unto itself, it tends to lose the notion of *meaning*. Meaning is a shared community awareness, often symbolic or mythical. Most often, our mind and our feelings work to unify sensory data, cross-referencing our images and our experiences. John Ashbery made a strong case for using narrative. He said simply that an art constructed according to singular canons or formulas will wither away since "having left one or more of the faculties out of account, it will eventually lose the attention of the other."[1]

Actually, the story has been nestled in art since people began drawing. If we return to the bison on the walls of the caves at Lascaux, France, circa 35,000 B.C.E., we will find picture stories about the hunt. One of the greatest illustrations in the world is Michelangelo's Sistine Ceiling, a giant tableaux depicting the Book of Genesis. Some contemporary artists are returning to narrative art. Consider John Baldessari and Roger Brown.

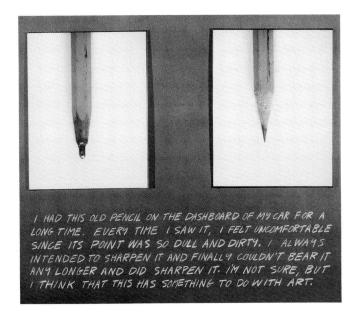

**Figure V.1** *John Baldessari.* The Pencil Story. *1972–73. Two Type R Prints on Board, 22 × 27 1/4 inches. (Seen in* Art in America, *May, 1981.)*

Baldessari places two photographs of a pencil on one panel with his own casual story line printed beneath. The picture represents his tongue-in-cheek allegory for expected artistic norms. Standards are what Duchamp (and hundreds of other artists) have usually fought against. In our culture, pencils are *expected* to be sharpened. Art is *expected* to be whatever artistic taste rides the cultural crest. Figurative art, for instance, gained respect again in the 1980s, but not until a period of general neglect preceded. So too, narrative work.

Roger Brown, in *Giotto and His Friends: Getting Even*, uses twelve contiguous panels, reminiscent of early altarpiece predellas (small story panels beneath a larger painting), to tell the tale of the Fourteenth Century Italian genius, Giotto, as if Chicago art critics had lived in Giotto's time. He tells of those art critics as provincial commentators and self-appointed taste setters.

Pictorially, Brown's work is reminiscent of early illuminated manuscript space: foreground images, vertically receding depth, and repetitiously stylized figures, buildings, and landscapes.

Many more than these two contemporary artists use narratives in their art works, but we need now to consider what will guide you to expressive work in this paradigm. *Figure: The Fulfiller of Form* has to do with the symbiotic, meaning "close," relationship of word and image. It assumes that something larger than either alone can result. For the artist to find that symbiosis, or closeness between word and image, several notions arise. The first is that language needs to function less as information and more as a gesture that infers a feeling or attitude. The second is that we know that images are always preferred to words even though both serve a larger continuity or reference. And the third notion is that narrative artworks are often assembled in pieces or segments. "Bit by bit," as the Stephen Sondheim lyric puts it.

Because a kind of thinking and construction applies to *Figure: The Fulfiller of Form*, much of the discussion of this paradigm will follow patterned processes. (This chapter does, by the way, provide an extended overview of most of the analytical terms artists use, setting them alongside a few literary terms.) As in all the paradigms, good craftsmanship is important. Master the building blocks, so that expressive concerns happen naturally in a dialogue between thinking and doing. Remember, talk to your stroke and your stroke will talk to you.

Let's begin with an outline of literary forms and what the artist needs from them. We will then move to a more extensive visual and graphic layout intended to do two apparently opposing things: amplify and simplify your choices. As you read the following paragraphs, you may reference the chart.

## A GENERALIZED OUTLINE OF ALTERNATIVES

1. The artist chooses a *Literary Form:* the *narrative, history,* or *biography.* If you should write your own storyline, you may complicate efforts by wandering away from the visual arts into fiction, journalism, or some other complex art form. Be advised good writing is difficult, although any personal prose or verse that prompts your imagination can be invaluable.

2. The artist chooses the *Narrative Subject Matter.* For our purposes, that choice has to do with anything good or bad that relates to the body, mind, spirit, or feeling of a human being or of a community. Because you are dealing with invisible concerns—motivations, dreams, fears, etc.—as well as visible ones, the first choices can be among the most difficult. Even very good writers have trouble deciding where their stories begin. Further, narrative subject matter can be tragic, comedic, romantic, or nightmarish, to mention just a few of the genres. A chicken-and-egg situation sometimes creates a block, as the student chooses either the subject matter, which then must have a literary form, or a literary form which employs intriguing subject matter. That's why a good starting point is often borrowed prose, whether from a master storyteller or from a lover's letter.

3. After reading the *Story,* the student describes his/her *sense* about the story, jotting ideas and key words on a sheet of paper. How could you interpret a portion of the story to illuminate a larger continuity? Will you draw the characters in their setting and scene with *humor* or *seriousness*? (Looking up words associated with comedy/tragedy will give you gradations between the two polarities.) For you, what inherent tendencies are suggested?

   The *Time Frame* you choose for your work has more to do with *continuity* than with real time or the time line of the story. Each of us knows that an author can write ten pages about a character rising from a chair and going to the door. So, too, an artist can use one, two, or more panels to represent the same sequence.

   Continuity of time is measured in several ways pictorially. *Immediate* journalistic moment is one. Excerpting the one slot of movement that carries the most story information could be drawn with quick, impressionistic strokes to support a brush-by action. *Instant replay* recaptures any movement, even in a sequence that indicates movement in reverse. *Slow motion* speaks to the implied speed of the figure's motion within the panel or panels. What characteristics could indicate the difference between instant replay and slow motion? A few clues here might help. Instant replay panels could be read from right to left after you have invited the viewer to read your panels uncustomarily backwards. A change of scale and color for a right-to-left replay could move in a series from a large drawing in color to a black-and-white the size of a newspaper clipping. The

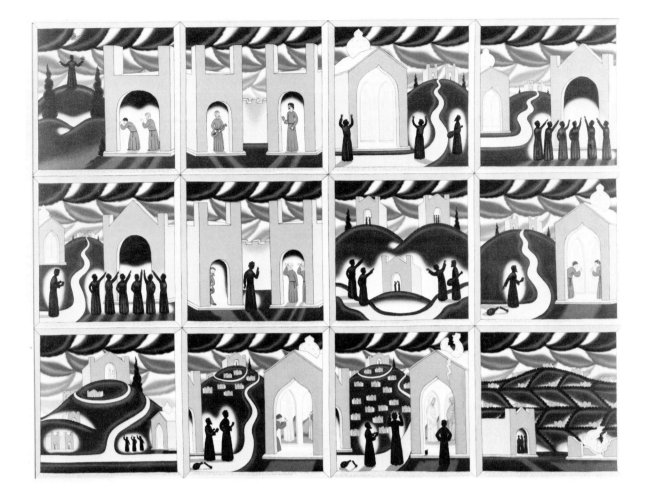

1. In Giotto's time there didn't exist an abundance of provincial art critics mourning the passing of the abstract conceptual style. Had there been a scene in Florence similiar to the one in Chicago, however, the story of Giotto might have gone something like this: Giotto and his colleagues were students of St. Francis of Whitney and Ray Cimabue. These young artists had grown weary of the lifeless Byzantine art of Italy's east coast and these great men stimulated them to rely on their own roots and individuality.

2. Now about that time in one of the cloisters were two effete monks, one a tall gaunt man with dark hair, mustache and black horn rim glasses, had earlier tried painting but finding himself totally without talent decided to be an advisor to others on the visual arts. The other was a lute player, fair of hair and sported a full beard. He fancied himself a person of great musical ability.

3. At this time most of the young artists of Florence followed the abstract style imported from the East. The lute player would often wander by while these artists bullshat about the wonders of the abstract style. He was fascinated by their word games and mental crap about abstraction. He was known to be holding out for the invention of a black box that would make pictures and would do away with all this silly hand made stuff.

4. Meanwhile, Giotto and his friends had created strong new work sharing a rejection of the old ways. They no longer looked to Ravenna and Constantinople to see what to do next. Their art had a human interest, visual richness, and accessibility not seen until this day. Many patrons commissioned new works by these painters and the people came in flocks to see them.

5. One day the tall gaunt monk noticed a great number of these new commissions by some of the most important art patrons of Florence. On looking closer he was amazed to find in Florence a very bold and innovative departure from eastern abstraction. This was unusual for the monk to see for it rarely dawns on the true provincial to look for art at home. They always assume real art is always made elsewhere and the next best thing is an imitation.

6. He sped back to the cloisters to spread the word. Some curates listened with curiosity and reluctantly had to agree that the new art was very strong and popular. The lute player, listening in the shadows, felt very threatened by this new art which was strongly visual, a characteristic he couldn't understand. As the word spread most curates across the country became excited about the new direction.

7. By this time the abstract artists, imitators of the East in Florence, were also feeling threatened, for the new art was spreading through the countryside. The lute player and the imitators spent many hours consoling each other on this turn of events.

8. The lute player having decided he was never going to make it as a musical advisor threw down his lute, took up pad and stylus and proceeded to spew out his impotent venom in the area of the visual arts where he had even less sensibility than he had in music.

9. But the new art had made an impression all the way to Rome. The curates, advisors, and patrons of Rome were all excited by the new direction. The adherents to Byzantine abstraction shook their fists in anger until the tall gaunt monk quivered in his sandals. He had initially spread the word and the lute player was nipping at his heels while the Byzantines were angry as hell.

10. So with the lute player by his side he went to see what the imitators of the East were doing and Behold! They were imitating Giotto! The tall gaunt monk and the lute player praised their work anyway as if they were great innovators.

11. Alas, they were all too late. The new art was spreading like wildfire. The tall gaunt monk threw up his hands in total confusion. He constantly offered contradictory advice. Being totally non-visual, the lute player continued praising the abstract style and babbling about pictures from a black box. The tall gaunt monk's advice began to sound humdrum like he was using a formula. He would always say something good and then immediately reverse himself by saying something bad about the same subjects. He became a laughing stock of Giotto and his friends.

12. Giotto and his friends grew in fame and fortune despite the provincial efforts of the two monks to degrade their work with snappy copy. The monks succeeded only in making themselves look more provincial than ever. As Giotto and his friends grew old and died the monks would praise their names as if they had never said a nasty word against them. Justice prevails, however, for the name of Giotto is remembered forever, but no one remembers the name of the lute player or the tall gaunt monk.

*Figure V.2*   *Roger Brown.* Giotto and His Friends: Getting Even. *1981. Oil on canvas 72 × 96⅜ inches. Photo courtesy of the Phyllis Kind Galleries, Chicago and New York. Photo credit William H. Bengtson. Private collection.*

strokes themselves could be impressionistic, nearly dissolving the figure for either continuity replay or slow motion.

The immediate moment, instant replay, and slow motion all suggest action or movement.

*Timelessness* typically refers to an archetype or prototype such as one might experience in a drawing of "the warrior of all wars," for instance. *Compressed time* suggests a series of events happening within one panel, often with some sense of motion. But, compressed time can be static. Static time has an airlessness about it. Static time has no motion. It is at rest, quiescent, even when a number of scenes are placed together.

Another type of continuity could be called *sequential*. Time and story line can be seen in illustrations sequentially when the picture predominates and selected scenes are used linearly from the beginning through the middle to the end of the story. For variation, the sequence may track backwards.

No doubt there are other types of continuity. And part of the power of continuity is its emphasis on inevitability. Or, on sudden discontinuity. Your visual imagination may lead you to draw moments unwritten in the original story.

The scene you choose and the time continuity you opt for in your drawing or drawings should indicate to you what lighting you need. *Natural lighting* (daylight, moonlight, lightning, water reflections) is one option. *Man-made lighting* (bulbs, candles, explosions, mirrored reflections) is another. *Luminous* light is the artist's concoction, as is *holy light*. In the fifteenth century light became a factor separate from color and shapes. Much unusual illumination originated in its use with biblical subjects, hence, holy light.

4. *The Presentation* comprises everything, all the instruments you will need to create a picture. Here the choices become numerous, detailed, and practical. The components include virtually everything you have studied heretofore.

In review: the *Subject Matter* is taken from the story—what *scenes*, what *character or characters*, what *setting*, and of course, what *lighting* (natural, man-made, luminous, or holy).

The *Materials* are chosen from wet base (inks, watercolors), dry base (pencils, pastels) or grease base (crayons, oil sticks). Any drawing support (paper, canvas, wood) would also be among the physical materials that help identify the final visual form.

The *Formal Elements* include the *point of view (p.o.v.)* meaning one-, two-, or three-point in linear perspective, a roving perspective, aerial perspective, overlap, or multiple points-of-view. The question is, where will you place the viewer?

*Lines* have different qualifications. The five categories listed here are not exhaustive, but include most line usages. *Contour* is stable, steady, usually showing the external shape or the topography of what you draw. *Gesture* is often used for quick studies to indicate the characteristics, the kinetic structure of the figure you see. An additional category might include some of both contour and gesture with varying degrees of pressure, width, length, and energy within the lines. *Cross-hatching* is a type unto itself with lines crisscrossing over each other to create volumes. *Mechanical* line is used very infrequently with figurative images, but some artists do make use of the ruled line.

*Values* relate to shading and shadows. They range from white to black with gradations in between. Six categories of lighting are generally assumed with natural and man-made lighting: highlight, light, shadow, core of the shadow, reflected light, and cast shadow.

*Colors* are individual hues, such as red, yellow, and blue.

*Textures* range from rough to smooth. "Visual texture" is something you draw. "Actual texture" is that of the paper (called the tooth). Actual texture can also mean the imposition of materials on top of the paper.

*Shapes* derive from positive structures and their negative displacements, from masses of lines, from wide lines, from value, color, and texture. They can be transparent, opaque, or stages in between.

*Space* refers to linear perspective, aerial perspective, simultaneous space, equivocal space, overlapping shapes, flattened space, and the uses of value. Space for an artist means working with illusion.

*Composition* is the organization of all the formal elements into a cohesive unit.

*Mode* refers to the kinds of strokes you will use. Usually, one has a *classical* approach with drawn shapes most often seen as specific, identifiable, and "closed" with contour lines and values. One uses a *romantic* approach with shapes that quickly capture essential characteristics. Those shapes most often are seen as gestural, suggested, inferred, or "open." Value is also used within romantic traditions.

5. The *Visual Form*, meaning the physical form of the work, ranges from drawings to game cards, from comics to border edges, from pillow books to murals.

6. *Scale* is always a critical choice. Miniature to life-size and ranges in-between are options. Where scale exceeds your studio, a "mock-up" or small version of the large piece could work.

The expressive question we need to raise is: What are you doing with your "presentation," with the whole mix of your subject matter, materials, formal elements, mode, and visual form? Does your presentation affectively support your perception of the story? After this overview of terms and tools, we are now ready to explore expression in narrative.

Remember, we are moving from a literary form to a visual form. Simple play with words is frequently the first and natural step into this paradigm. Using two characters from the children's story, "Red Riding Hood," as an example, let's see how we might draw images that start with words. For instance, what comes to mind when you describe the title character? Perhaps young, innocent, caring, healthy, sweet, and naive might make up your list. Write them down, or write others.

Now, using formal elements for each of those words, visualize what the words "look like." What kinds of lines, values, colors, textures, shapes, and space would indicate "young"? Your associations are important, and you might try "free association" by speaking your words aloud. Predominant words, repetitions, or puns might surface. This word play will eventually lead to the choices that make up your image.

For "young," lines might be narrow, thin, and gestural. Values could be light tones, modeled to soften edges. Soft colors, perhaps pastel blues, greens, and yellows, are suggested. Chances are textures would be refined. Shapes might move to rounded forms. And more than likely, space would entail a logical, progressive perspective, idealized. "Innocent" carries similar connotations. "Caring" does too.

The word "healthy" takes on other connotations. Vertical, wider, bolder lines are feasible. Workable values move from middle to deep-middle tones. Colors probably are stronger. Textures might show contrasts. Shapes suggest more specificity. Space still infers realism, somewhat idealized. "Sweet" and "naive" are similar enough to "innocent" and "caring" that we probably do not need more descriptions.

Visually, we have some elements we can work with now to draw Red Riding Hood. Soft lines, light values and colors, idealized shapes, and space are suggestions consistently surfacing, all probably prompting feelings of vulnerability, the kinds of feelings that start to be expressive of the inherent drama.

What about the Big Bad Wolf? What words describe him? Big, bad, ugly, hairy, fierce, hungry, cunning, a conniving creature, are all possibilities. Using again your associations of words to formal elements, what seems likely to surface are visual alternatives to Red Riding Hood.

"Big" indicates both wide and heavy lines, both vertical and diagonal lines (diagonals are our most irritating line because they are neither moving up nor falling down). Values and colors would likely be dark and heavy. Shapes would seem to have obvious edges. Comparative scale becomes important. "Big" is always relative to "small."

"Ugly" suggests jagged lines, intense values and colors, very rough textures, and identifiable, coarse shapes.

"Hairy" infers dense lines, heavy at one end, lighter at the other, with sharpness in the stroke. Dark against light values would be apt. Colors refer to extremes, black and white. Bold shapes with matted textures could be used. Space feels unkempt.

"Fierce" could use heavy, jagged lines, very dramatic lighting (whites next to blacks), biting colors like neon greens and pinks, and irritating textures. Options include partially understood shapes fading from abrupt strokes to wisps. Space would seem to be dense, dark and mysterious. Perhaps eyes peering from other dark shapes might be useful in supporting the expressive content of "Wolf."

"Hungry" indicates thin, nervous, meandering lines, black ink on white ground. Eerie lighting is suggested. Profuse textures apply to nearly all shapes. Shapes are thin and tight. Space appears vertical, theatrical.

"Cunning" suggests short, abrupt, horizontal lines, grey, black, white, and purple colors, mysterious lighting (perhaps back lighting) close harmony in values (for instance, middle tones on top of middle tones and darks beside darks.) Textures could be physical through the application of media onto paper. Shapes appear to be stretched, taut, horizontal without much mass. Space would be indeterminant, hard to understand, perhaps by scrambling one-, two- and three-point perspectives in the same scene. Or, space could be abstracted in darkened tones.

The word associations for this children's story are the typical ones. You might select others, and push the narrative to extremes. Your expressive task will usually include surprise, the shock of reality, and the interesting disclosure of the narrator's visual point-of-view. Action comics, as you may have observed, often place the viewer at angles oblique to the subject matter, sometimes demanding sudden perspective shifts from panel to panel. And remember, in addition to character traits in a story are character actions: the journey through a dark place, the losing of innocence, the rescue.

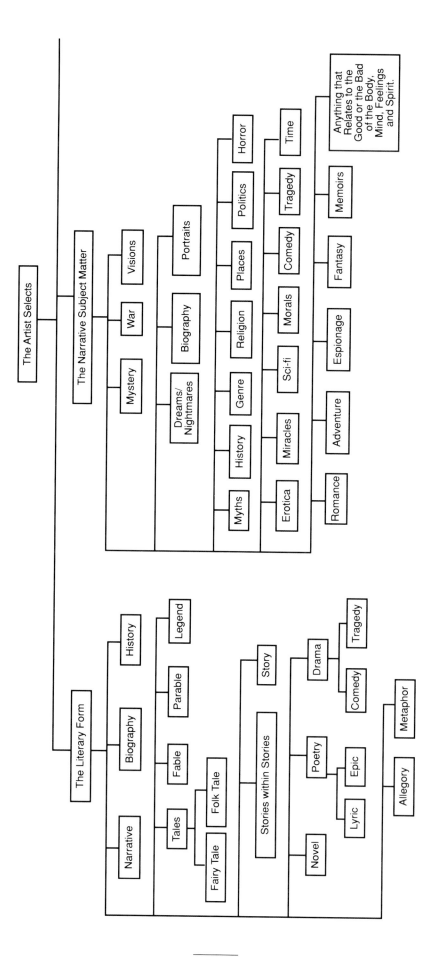

The Artist Selects

**The Narrative Subject Matter**

- Mystery
- War
- Visions
- Dreams/Nightmares
- Biography
- Portraits
- Myths
- History
- Genre
- Religion
- Places
- Politics
- Horror
- Erotica
- Miracles
- Sci-fi
- Morals
- Comedy
- Tragedy
- Time
- Romance
- Adventure
- Espionage
- Fantasy
- Memoirs

Anything that Relates to the Good or the Bad of the Body, Mind, Feelings and Spirit.

**The Literary Form**

- Narrative
- Biography
- History
- Tales
- Fable
- Parable
- Legend
- Fairy Tale
- Folk Tale
- Stories within Stories
- Story
- Novel
- Poetry
- Drama
- Lyric
- Epic
- Comedy
- Tragedy
- Allegory
- Metaphor

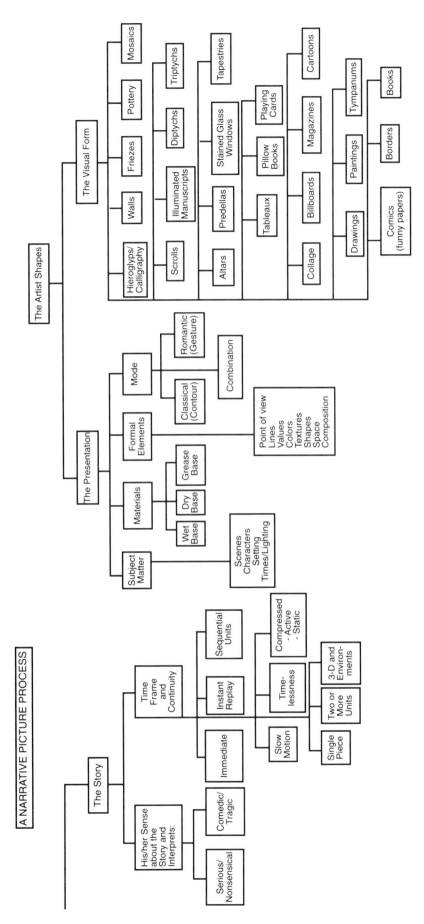

**Figure V.3** *The Picture Form Chart.*

By now you have a few more clues to put your images together. Using narrative components alongside visual elements can move you to draw images in ways you might never have imagined or tried.

Increasingly, your decisions may seem to come automatically. Difficult first choices lead to easier and more inevitable second and third choices. And you will find, almost as a by-product, that awareness of narrative picture building also helps you understand why something is not working in other figurative, nonnarrative work, especially when you begin to feel stuck in the process.

While you are looking at examples in this paradigm, take note of what each artist chose in support of the content of the narratives. Ask yourself what you would have done differently.

Words in and of themselves are not stories. They make reference to their own meaning. But some artists can use a word or two, which, when given a juxtaposing image, imply a new content for the word. Such is Ed Ruscha's "Cut." The image implies blood by the way it lies on the paper, the lighting used and the singular, centering of the word to demand our painful response: once we are cut, we cannot concentrate on anything else. These are associations and do not in themselves make a story.

But words can be used as part of an illustration. Our student work incorporates calligraphy and design in her book illustration of Narcissus from Ovid's epic poem *Metamorphoses*.

Terry Allen's *Studies for the object/drama . . . WHISPER*, is an eight panel series, a full storyboard for a drama that has no other form apart from the drawings in frames. Echoed male and female shapes are silhouetted, heads together, in the foreground. The word "Psssst" is repeated around the image. The other phrase, " . . . the modern era as a possible explanation for the modern error," alludes to the frustration that viewers now must read and not see art works, such that viewers must take a speed reading course just to view pictures. We are taken into the picture by being placed behind figures who stand before the front of a benign neighborhood house, and like a voyeur, we can smile about the comment on taste made by the silhouette characters. This is a narrative experience. The experience would be more fully developed were we able to include all eight panels.

Allen's *Literary Form* is the *allegory*, a symbolic representation. His *Narrative Subject Matter* is a chronicle of artistic taste. His *sense* about his subject matter is *tongue-in-cheek/humorous*. His *time frame* in our illustration is one day.

**Figure V.4**  *Edward Rushca. Cut. 12969. 36" high. Iolas Gallery, New York.*

**Figure V.5**  *Jeri Peterson. Narcissus. From Ovid's Metamorphoses. Ink and watercolor, 22" × 14".*

The *continuity* in the whole piece is within eight panels. We have just one piece illustrating our points. For presentation his *media/materials* are pencil, watercolor, typeface, and ink on paper. The *formal elements* include a one-point perspective (perhaps an allusion to the public's conventional view of what art ought to be), lines in the roof, sidewalk, architecture, bush, trees, and silhouettes. Values range from white to black. Textures are developed through the use of lines, flat images, and shading. The shapes of the silhouetted figures are very prominent. Space is flat and frontal, but moves only into a middle distance. The composition keeps one's eye shifting from figure to figure, occasionally glancing up the sidewalk to the door of the house. The *Mode* is classical, with identifiable, closed forms.

The *Visual Form* is a series of framed drawings, ongoing boxed commentaries on public taste. We see only one in this example.

We will now examine several stories, beginning briefly with the Old Testament story of *Cain and Abel*. Cain and Abel were the two sons of the first family of the Bible, Adam and Eve. Cain, the eldest, grew to be a farmer, but when he brought grain for a sacrifice, he displeased God. Abel,

the younger, grew to be a herdsman. He sacrificed a lamb, which pleased God. Cain, in jealous anger, killed his brother, Abel. Cain was sent by God to wander for the rest of his life.

William Blake has chosen the *Literary Form*, the legend. The *Narrative Subject Matter* is religious. Blake's sense of *The Story* is serious. His *time frame* continuity is protracted but in a condensed way, meaning, that there is a sense in the drawing that several events culminate into one pictorial image. Cain dug the grave for Abel's body, Adam and Eve found his body, Eve is grieving over his death, Adam is shocked and grieving both his eldest son's sin and departure, and the dark spirit of wrath and fury are forcing Cain to flee.

Blake's *Presentation* uses *formal elements* that include a one-point perspective used to intensify feelings in this image. Horizontal lines of shovel, grave, and dead body, are disrupted with jagged lines of mountains, and strong diagonals of Adam's arms/hands and Cain's fleeing, stretched, taunt, thrusting body. The short, abrupt, nervous, flame-like line surrounding Cain suggest he is fleeing in terror.

***Figure V.6*** *Terry Allen.* Studies for the Object Drama . . . "WHISPER". *23 1/2 × 29 1/2", enamel water color, type, ink, pencil on paper. Morgan Gallery.*

**Figure V.7**  *William Blake*. The Body of Abel found by Adam and Eve. Cain Fleeing. *Circa 1826–1827. Tempera on wood, 12 13/16" × 17 1/16". Tate Gallery/Art Resource, N.Y.*

Color is another element used to support the story line. The bodies and sky are deep gold. The grave, mountains, cloud, and "terror lines" are blackish in hue. The huge blood-red sun hangs in the middle of the work separating spirit-cloud, Cain, and his parents.

The figures are large in scale before the mountains, probably to intensify the drama of the scene. Although the grave lines seem sharp and the mountains jagged, there is a softness to the edges of the shapes.

The space used is similar to a stage space with set designs placed behind the actors. This unreal space heightens the paradox of the murderous act, distanced participation. The figures remain in the foreground.

The composition is held tightly together. The eye cycles continually from Cain to fire-red sun to cloud to parents to Abel's body to the grave to Cain's foot and up around again throughout the composition. The movement feels unremitting, like the results of a murder. Emotional responses resulting from a murder also cycle again and again.

The *Visual Form* is a small tempera work on wood, one panel, a tightly compacted, intimate form symbolizing dense layers of human experience.

## A Study:

Another example of compressed time is seen in Linda Baechler's illustration of *Alice in Wonderland.* You remember the storyline. On the bank of a small river, Alice, beside her sister, falls asleep and dreams she is following the White Rabbit down and down a hole into a dreamland of characters and companions. The Mad Hatter, the March Hare, the Doormouse, The Cheshire Cat (who, despite his vanishing trick always leaves his grin behind), and, of course, the outrageous Queen of Hearts with her retinue of nervous playing cards are among characters that Alice meets. Baechler was more selective than author Lewis Carroll.

The *Literary Form* is fantasy. The *Narrative Subject Matter* is dreams. The artist's sense about *The Story* is playful. The illogical sense of time is supported in *continuity* by using compressed time simulating the appearance of many characters simultaneously. Within *The Presentation, subject matter* includes the hole through which Alice fell, Alice, the Rabbit, Tweedle Dee and Tweedle Dum, the Mad Hatter, and sundry others. The setting is less a setting than a montage of clouds and grass with the characters interwoven. The lighting is made up, having less to do with a lighting source than with a simple shifting of values; she selected and moved lights and darks variably around the composition.

**Figure V.8**  *Linda Baechler.* Alice in Wonderland. *Watercolor, ink and pencil, 18″ × 24″.*

The *materials* are pencil and watercolor. The predominating *formal elements* include line, which is used sparingly to suggest a full figurative character. Both values and colors infer much of the rest of the figures that line has not. Values and colors in this work lend themselves to the more playful and nonthreatening characters. The Queen of Hearts, for instance, can be intimidating, to say the least, and she is not in this work. Textures are limited. Shapes often are a composite of color and value overlays with hair becoming grass, the rabbit transformed into cloud, which of course means space remains dream-like. The composition is almost circular. The "drop entrance" to the cave in the upper left area takes one to Alice, then to her foot, which points the way through the rest of the composition.

The *Visual Form* is a one-unit watercolor painting/ drawing, allowing a viewer the same sense of freedom one feels reading the story.

In contrast, Gabriel Laderman's large polyptych, *House of Death and Life,* is a good example of compressed time in stasis. The people are motionless, static.

Laderman uses a short story as the *Literary Form* from which to build his doll house units. The *Narrative Subject Matter* is mystery. The *time frame* is compressed and static.

The whole piece is made up of six pieces, a polyptych. The artist's *sense about The Story* is serious, unemotional tragedy. The *Presentation* includes figurative *subject matter* within rooms in a house. There are six interdependent scenes, which with one exception, include people. The setting is an interior. Because lights are on inside the house, time appears to be night. The *materials* are oil on canvas.

The *formal elements* incorporate incongruent one-point perspectives within the rooms and between each room scene. At the same time we are placed just above the rooms, requiring a low three-point perspective. Purposefully displaced perspective leaves an uneasy feeling. The p.o.v., is where the artist places us, the viewers, and in this case we do not know where we stand. Because we are unsure now, a denominator of anxiety is raised with this picture. Some of the figures are drawn as though we are looking at them (one-point) while others are drawn as though we are looking down on them (three-point). The scenes thus indicate the kind of disjointed space that can sustain the enigma of the subject.

Lines tend to be contour with stiff, closed shapes. Nearly every item, including the people, is finished with hard-edged lines. Values are dramatic simply because the lighting is. If one looks just at the shapes of lighting one sees an intricate pattern in the scenes of the lower half that swoop up to the more rectangular light patterns above, so that one then looks at the strange goings-on of the figures. Rooms are brightly lit below, dimly lit above. Space is the interior of a number of rooms in a house.

Because of the closed shapes, the stark lighting and the noncommunicative body postures of the characters, a sense of removed intensity supports the mystery of the portrayal.

The story here loses a linear construction. In the upper left room a woman is being strangled by another woman. The seated figure in the room below appears to hear noises and is looking up. One woman, smaller in scale, stands with one foot on the stairs. Two seated women, one clothed, one nude, in the kitchen on the right do not face each other. The figure lying in the upper right room faces a nude

apparition within the wallpaper. Narrative continuity becomes a puzzlement.

The *Visual Form* is quite a unique construction. The whole piece measures 93″ × 135″, which is, of course, about 7′7″ high, by 11′ wide. This composite work has four large outer panels and two small central panels, arranged like a two-decker, cut-away doll's house. A line of tiles crosses the rooftop. Trompe-l'oeil cornices at the tops of each room (painted to look realistic) sustain the sense of a doll's house paradoxically removed but still fascinatingly mysterious. The static sense of time and the pervasive immobility of everyone and everything makes one think of an archaeological dig. The layers of meaning herein need sorting out. The mystery needs to be solved.

Another type of story, the myth, often a common denominator for many people between religion and poetry, is summarized by Marcia Moore in a Hopi Origin Myth.

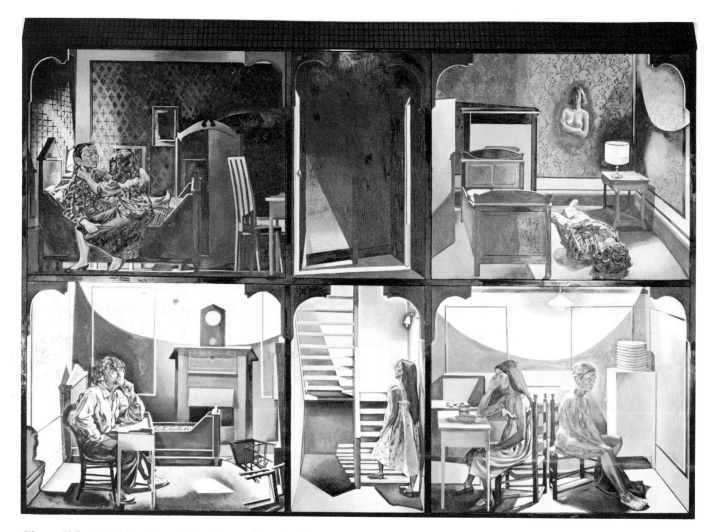

***Figure V.9*** *Gabriel Laderman.* The House of Death and Life. *1984–85. O/C, 93″ × 135″. Schoelkopf Gallery, New York. [Seen in* Art in America, *April, 1986, page 172.]*

## A Study:

In the Beginning were three universes, Fire, Air and Water. They were of no use separated, so the Creator destroyed them. Survivors found themselves in a raft on an ocean. The survivors were lifted up through a "sipapu," a hole in the center of a sacred chamber. Mother Earth and the people become the fourth universe.

Because the number "four" is important to the Hopi for what it symbolizes—four directions, four divisions of time and of life (babyhood, childhood, adulthood and old age), and four things above the world (sun, moon, stars and sky) the student drew the survivors entering the "sipapu" with a woman, Mother Earth, standing by. Four people emerge and face East because, according to belief, life rises there.

The *Literary Form* is the legend. The *Narrative Subject Matter* is the myth. The artist's *sense about The Story* is serious. The *time frame* is timelessness. *The Presentation* includes subject matter with characters in three scenes. The setting is of a refined nature with figures moving through a hollow log. The time is "anytime." The lighting tends to come from right to left (East to West perhaps.) *Materials* include colored pencils on rag paper. The *formal elements* begin with the same point of view for each scene unit. Most of the artwork is done in contour line, then filled with colors and values. Textures are limited to repetitions of line simulating ground or cloth. Nearly all of the shapes are taken from nature, but with the kind of line and shapes chosen, along with the flattened colors and values, any sense of depth comes through the topographical contouring in line. Space is viewed by way of a relatively low three-point perspective. We are looking down on this group. The composition moves from left to right, with an abrupt "zoom in" on the second panel. However, our eye stays within the unit because of lines and values that counter the motion of the figures moving right. The visual mode is classical, the people and objects are closed shapes and identifiable. The *Visual Form* is not unlike a triptych but with two panels outlined, the third left "open-ended," similar to the journeys each of us makes from birth through life, physically, psychologically and spiritually.

A literary fable is a concise narrative making a point (often a "moral"), frequently using animals that speak, act, and think like human beings.

## A Study:

Karen Burkwall used Rudyard Kipling's "The Cat That Walked by Himself" from the *Just So Stories*. Paraphrased, the fable goes like this: when the tame animals were wild, the wildest of all was Cat. Man was wild too, but was tamed when he met Woman. Man and Woman had a child and lived in a cave. One day Dog smelled mutton cooking and went to the cave. He promised to guard the cave in exchange for the food.

One by one, the wild animals exchanged freedom for food. Wild Horse promised to wear a bridle in exchange for new mown hay. And wild Cow promised to give up her warm milk for grass. Wild Cat was more shrewd.

Wild Cat heard that Baby liked soft, fuzzy things with which to play. So, Cat went to the cave and played with Baby when he cried. The Woman blessed the cat. Again, Baby cried and Cat told Woman to take a thread and run it over the ground. Cat chased the thread. Baby laughed and fell asleep. Woman blessed Cat. The third thing Cat did was to catch a mouse that frightened Woman. So Woman promised wild Cat he could sit by the fire in the cave and drink warm

***Figure V.10*** *Marcia Moore.* The Hopi Origin Myth. *1983. Colored pencil on rag paper, 15″ × 30″.*

white milk three times a day for always. He is Cat who always walks by himself, waving his wild tail and walking by his "wild lone."

*Literary Form* is fable. *Narrative Subject Matter* is morality—the moral of the story is—find someone's need, fill it, and retain your own integrity. The student's *sense about The Story* was humorously serious. The *time frame* derives from chronologically sequential units in that excerpted parts of the story illustrate the action from beginning to end.

*Presentation* includes the *subject matter* of animals, people, and nature, mostly in daylight. The *materials* are rag paper, colored pencil, and graphite. Nearly all *formal elements* are incorporated. The p.o.v. changes with different scenes. We look directly, then up and down at the figurative images. All positions are related to linear perspective. Apart from the use of contour, lines throughout help shape several different formats. Two panels use an arched outline, simulating the roundness of the cave roof. And two panels

are used for a singular illustration of Wild Cat in his autonomy at the end. Values follow natural dispositions, though a specific light source is not always clear. Colors are vivid, varied, naturalistic. Textures are limited somewhat because the scale of this piece is small. Shapes follow the colors and values in that they represent objects. The cat is always shown off on the side or by himself. And space generally carries the illusion of recession. The *mode* is classical with line predominating in closed and understood shapes. A more gestural line, for instance, would not have seemed as appropriate as the contour for this moralistic fable and its characters.

The *Visual Form* is a small, intimate book, which accordions together for outside dimensions of 4″ × 6″. The end pieces are the simple front and back panels of the book. This little book is a very private experience. The book nearly begs you to be by yourself in your own "wild lone" to experience the story.

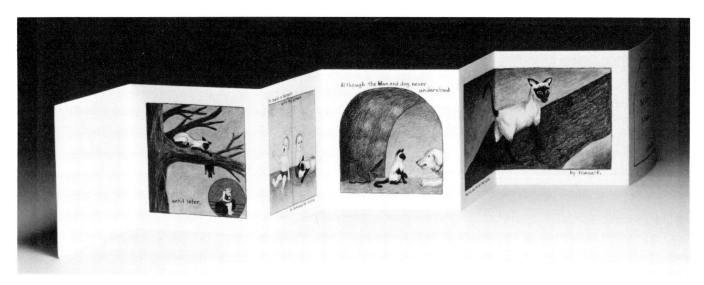

***Figure V.11*** *Karen Burkwall.* The Cat Who Walked by Himself. *By Rudyard Kipling. Colored pencil, graphite, rag paper, 4″ × 6″ closed.*

---

The next two story sequences involve *time* as subject matter. The first draws upon fleeting transitory changes in time and atmosphere. "Now you see it ( and not very well), now you don't." The second spots key transitional eras in art history. Jim Negley's work, *Tree and Fence,* uses the viewer as witness to a single figure as it is effected by the movement of the landscape and atmosphere that forces of nature destroy and rebuild.

Before analyzing further we might reference (again) T. S. Elliott's "objective correlative," and ask at what point does the abstraction of an object, a figure for instance, become something other than the figure? When does the figure, the subject matter, no longer express the model? When is the image no longer correlated to its objective starting point?

## A Study:

Negley's story line, as such, appears to be the growth transformations of the figure-as-monolith. We begin with a small monolith on the book's cover. Ensuing pictures show the monolith on top of a female torso. Tree and fence become part of the monolith. The monolith splits at one point, the figure begins to dissolve and tree, fence and monolith take over. About the 11th, 12th and 13th page, the figure loses its correlate and we can no longer identify it as figure.

Applying our narrative schema to this sophisticated and interesting piece, let us see what we come up with. The *Literary Form* would be poetic narrative. The *Narrative Subject Matter* would be "time," in a philosophic sense. The artist's *sense of The Story* seems to be humorous wisdom. The *time frame or continuity* is in 14 sequential units, which become a three-dimensional, bound book, about 24″ × 18″.

Subject matter for *The Presentation* is not full of characters, scenes, and settings *per se.* There is a monolith, a female torso, a tree and a fence within various landscapes and weather.

Negley shares his experience with the *materials* and the process so that others might try similar approaches. In the drawing process Jim began with two papers, the book pages, side by side. He quickly sketched out the figure and the primary shapes. All basic decisions of scale of figure to landscape were made early on. Once made, the pages were spray fixed.

Layers of acrylic washes were applied to the surface, the strokes loose and gestural. These layers helped establish the color scheme. Whatever choices were made, even weak ones, were left alone to be modified by later layering. He wanted principally to "break the whiteness" of the bare paper.

Next, a piece of rice paper was glued to the center of the drawing, covering parts or all of the figure. The size of the rice paper was important and was planned carefully. Following the rice paper was the drawing of the tree and fence with conte crayon and compressed charcoal. Another spray fix. Layers of pastels, conte, charcoal, and hints of acrylic paint were carefully applied in the final steps, fine tuning the forms, colors and values. Each page was compared to the page before and after to maintain continuity, yet retain a singular uniqueness. Bit by bit. When finished, the artist hand bound all pages and covers of the large book.

*Formal elements,* as such, are adjusted to fit the artist's needs as the work told him those needs. For instance, p.o.v. is usually associated with linear perspective. We do not have linear perspective in Negley's work. We are simply presented with images in foreground, middle ground and backgrounds, each of which have variable horizon lines. Lines indicate the outline form of items. As these lines become wider they signify the shape of things, such as the fence. The colors are in greens, blues, purples, blacks, and whites. Textures are visual and actual, especially by way of the imposition of rice papers onto the work. Space is defined largely with overlapping shapes, change of scale, atmospheric effects, and the continuity of landscape. Compositions, all fourteen, are vertical in format and are self-contained. There are some lines, such as those denoting cloth, snow, or horizon lines, that feed from one picture into the next.

The *Visual Form* is one oversized hand-bound book. The beauty of this story of continuity unfolds as the large pages are turned. Continuity is a time where nothing is permanent but change.

The next series transitions us into "history drawings," the drawings and paintings of historical moments and places.

The title of the first piece is *Ding Dong the Witch is Dead.* The title is an allusion to the twentieth century artist's need to find a "new reality" besides the verisimilitude represented in a Mona Lisa. Since Masaccio's works in 1420s in Florence, Italy, artists have been painting with a concern for accuracy by using one-, two-, and three-point perspectives. These practices transitioned through several schools: stretching the figure in Mannerism, for example, to painting bawdy genre studies in bars, to untouchable neoclassical figures, to all the other "isms" principled in the belief that truth in beauty had to rest in the illusion of natural space.

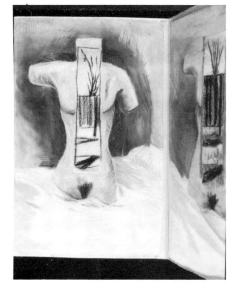
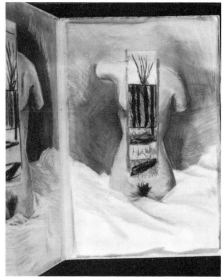

**Figure V.12**   *Jim Negley*. Tree and Fence. *1989. Mixed Media, bound book, 24″ × 18″.*

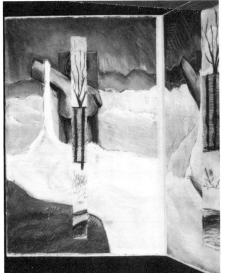

**Figure V.12** *Continued.*

## A Study:

Michon Weeks took the *Literary Form* of history, art history to be exact. The *Narrative Subject Matter* is the chronology of "time" in art history via the portrait of Mona Lisa. The *time frame or continuity* lies in forty-one sequential units. Her *sense about the historical line* is humorous. Subject matter in the *Presentation* is Leonardo's *Mona Lisa*. Among the *formal elements* the p.o.v. with "Mona" never changes. Lines remain with contour. Values and colors tend to change according to the stages of dress and undress shown. Shapes and space move from naturalistic forms to flat patterns to no pattern. The composition sustains the triangular unit of the three-quarter turned woman, arms crossed at her wrists, with variations. The materials are rag paper and colored pencils. The *mode* tends toward classical.

The *Visual Form* is one horizontal composite with forty-one 4″ × 3″ drawn units. Placed horizontally on a wall "Ding Dong" allows us a bird's eye view from left to right, of the transitions Mona takes from the Italian Renaissance in the sixteenth century through the Baroque, Rococco, Romantic, early Fauvist, Cubist, Abstract, Non-objective, even Minimal, art in the twentieth century.

Works of *Historical Narrative*, as in story narrative, originate with early human efforts. Egyptians made bas-reliefs of a ceremonial opening of a canal by the king, c. 3,000 B.C.E. Throughout the history of different cultures one can find visual documentaries of battles and sieges, the capture of law-breakers, the moments denouncing cowardice, the evidence of good and bad governments, the discoveries of new lands, the signing of deeds, the raising of flags, and so on.

Most history paintings and drawings were done largely before the advent of the camera. History, in whatever form, tends to be viewed through the eyes of the creator of the image, which often means that the end result is either greater than or less than the actual event, depending on the artist's perspective and belief. Many artists chose to paint the leaders or signers of declarations in heroic postures. Other artists had to paint heroically or lose their commissions. But remaining wholly objective and unbiased is difficult for even the most honest interpreter.

Thomas Eakins seems to have had an eye for objectivity. His *Gross Clinic* documents an operation in progress with stark realism, including all the blood and surgical details.

Eakins' *Literary Form* is history. The *Historical Subject Matter* was surgical medicine. Eakins' *sense about The Event* was serious. His *time frame* was a moment in the middle of surgery, one panel. *Subject matter* includes patient, doctors, clinicians, a woman, and observing students seated in a surgical amphitheater. The lighting is interior lighting, from left to right.

*Materials* in our illustration are India ink and wash on cardboard, a small preliminary work for the life-sized painting finished later.

*Formal elements* include a one-point perspective. Lines are indicated through brushwork that often separates lights from darks. The value scale shares an intense light over the operating area with progressively diminishing light the farther back the images appear to go. Textures are incorporated according to objects and figures, realistically. Shapes in the foreground are fully realized in volumes. Those figures behind and above the focal area are more suggested than completely finished. Space is within an interior lecture room and relies on the appearance of natural recession. The composition gives our attention to the patient and the doctor in several ways. Lines are more refined for details in the foreground. Lights and darks are most dramatic at the center of interest. The one-point perspective keeps us strongly focused. Body attitudes and psychological lines do the same. Two men are looking at the incision. One man looks down at the patient. The man at the bottom right looks up at the standing doctor whose eyes move to his right (our left), but whose hand and surgical knife point back to the table and the patient. The picture has a very focused center of interest relieved somewhat by the middle-toned desk just behind the standing figure. Behind the desk, taking notes, is a student whose body leans toward the doctor. The student's leaning body can also send the viewer's eyes up into the other seated students, but the lighting in the focal area brings one back down again.

The *mode* is a combination of classical and romantic, with much use of contour and gesture. The *Visual Form* is a wash drawing in stark black and white, not unlike scientific decisions made during operations.

Goya documented a terrible incident that happened before an audience in Madrid, Spain, June 15, 1801, when the gored body of the mayor of a small town hung on the horns of a bull that burst its way out of the ring and into the crowd.[2]

The *Literary Form* is history. Goya's sense of *The Story* was serious. The *time frame* is a stop action. *The Presentation subject matter* shows men and women, in panic, running for their lives, stumbling and falling on one another. A matador's bull is loose and has gored a small town mayor, a bystander. The action happens during the day.

The *materials* used were red chalk and paper. Goya's use of the *formal elements* include a p.o.v. about eye level. The *lines* are gestural, varying in thickness, density, and pressures to build toward the center of interest, the bull, which is a dark shape against a light background. *Textures* are seen less in this work as part of the composition and more as the rough tooth of the paper on which Goya

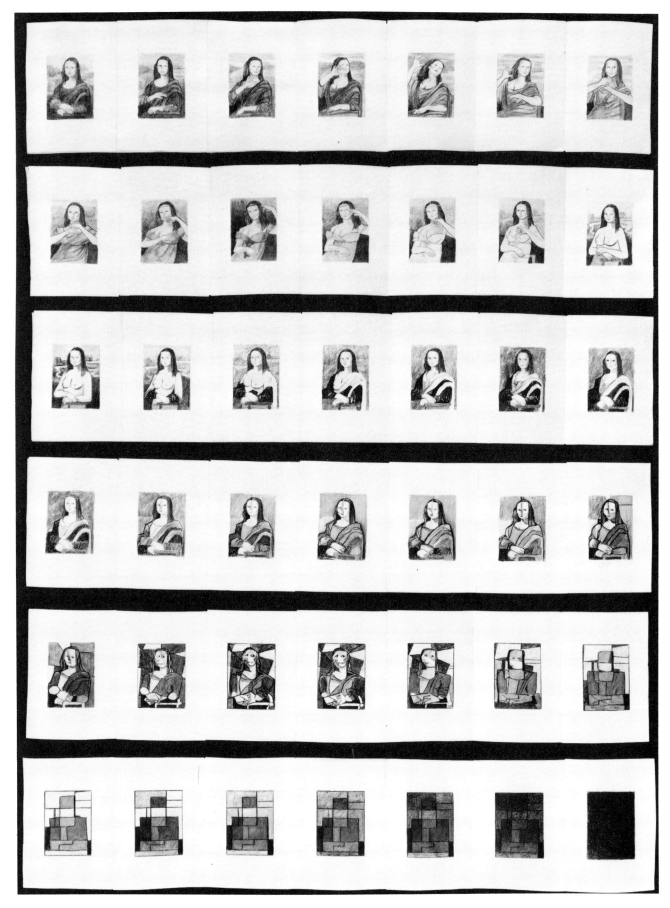

**Figure V.13**   *Michon Weeks.* Ding Dong the Witch is Dead. *1984. Colored pencil on rag paper.*

251

**Figure V.14**  *Thomas Eakins. Gross Clinic. c. 1875–1876. India ink, wash on cardboard, 23 5/8″ × 19 1/8″ (60.0 × 48.6 cm.), Metropolitan Museum of Art. New York, Rogers Fund, 1923.*

worked. Horizontal chain lines set up an eerie sense of figures being trapped. (Chain lines could also indicate the way the paper was made. Our contemporary charcoal papers often have lines similar to these that interrupt drawn images because they are a part of the paper, not the picture.) *Shapes* incorporate multiples of people piled on each other in an ascending pyramidic diagonal from left to right. The whole *composition* moves dramatically from left to right, with the eye ultimately lingering on the gored body of the mayor.

Because that sight is hard to look at for a long period of time, we move back to the left through the use of the long value shape of the taut body of the bull and his flickering tail. We cycle around through the figures, inevitably returning to the horror.

The *mode* is romantic, gestural. The *Visual Form* is one framed chalk drawing, one of a number of drawings Goya did of the bullfights. Seen together, figurative action is ongoing, as are the drawings in the Prado Museum in Madrid.

***Figure V.15*** *Francisco Goya (1746–1828).* Tauromaquia 21. *1815–1816. Red chalk. 185 × 300 cm. Prado Museum, Madrid, Spain.*

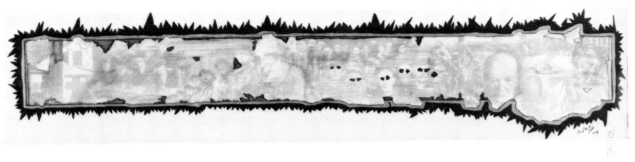

***Figure V.16*** *David Gale.* World War II. *History drawing,  9″ × 24″. Newspaper transfer and colored pencils.*

## A Study:

David Gale's illustration that follows deals with aspects of World War II. It shows less about a specific event than that event's mien in history. Prominent figures of the time, prisoners, buildings, and the Statue of Liberty, are transferred onto drawing paper from newspaper clippings.

The *Literary Form* is prismatic journalism. The artist selected World War II newspaper images as his *Narrative Subject Matter*. The *time frame* is compressed. The student's *sense about this History* is serious. *Presentation subject matter* does not have specific scenes, settings, time, or lighting. But the work does have faces and images of persons who were a part of an era. *The materials* are newspaper clippings transferred onto drawing paper with a transfer fluid. One does this by soaking the newspaper in fluid, then placing the newsprint picture face down onto the drawing paper. Rubbing the back of the newsprint picture with a smooth, hard surface, like a spoon, will transfer the image. Colored pencils were also used. Within the *formal elements* there is no specific p.o.v. On the left, we look up at a building, yet we look down on England's Winston Churchill on the right. The most dramatic line is the one surrounding the cannon-like shape of the drawing. The line is jagged, reminiscent both of grass and barbed wire. Values and colors vary according to the type of newsprint used. Sometimes the colors are enhanced with colored pencil. Textures are limited to the rubbing strokes required to make the transfer. The most dramatic shape is the singular outline of what appears to be a cannon surrounded with the grassy/barbed wire line. Within that shape rest the images of those who were part of the WWII experience. The composition is horizontal, long and narrow. We enter usually through Churchill's frontal gaze, but stay within the composition because of the repetitions of the blackened eyes, the "death eyes."

The *Visual Form* is one drawing, 9″ × 24″, reminiscent of a cannon, a object of death, and a long earthen grave, receptacle of dead bodies.

*Biography* is the third *Literary Form* we include in this paradigm. Biography generally is assumed to be portraiture, self-portraiture, parts of, excerpts from, or anything related to the facts or events of a person's life.

Saul Steinberg, a long-time cover artist for the *The New Yorker Magazine*, appeals to the intellect through humor and irony. Among the things that preoccupy Steinberg are history, geography, and biography. There is a "biography" of Millet. John Asbery, author of an article on Steinberg, has this to say: "Millet is an artist Steinberg likes because he was born exactly one century before him, in 1814. . . . Currently Steinberg is fascinated by the two figures in "The Angelus," and has had a rubber stamp made of them so he can place them in all sorts of unaccustomed environments, such as the beach at East Hampton or the desert at Gezeh. The biographies are actually certificates, certifying life by means of official seals and rubber stamps, portrait medallions and passages of handwriting that is illegible but looks as though it ought to be legible."[3]

The *Literary Form* is biography, rather allusions to biography. The *Narrative Subject Matter* is Jean Francois Millet. The *time frame* is understood through simulated documents. Steinberg's *sense about Millet's Biography* is humorous. *Presentation Subject Matter* is much like a certificate. There is no setting, just a series of seals, lines, calligraphic squiggles, repeated rubber stamp figures. Associative items remind us of the life of the artist. Lighting is undefined. The *materials* are paper, pen, and ink.

Some of the *formal elements* are used, but not in the traditional sense. Line and values are foremost. Values incorporate a range of lights and darks to sustain some sense of the illusion of recessive space. But the rubber stamp seal of the figures from Millet's painting titled "The Angelus," is pressed onto different backgrounds, like foreign deserts (perhaps an allusion to Millet's preoccupation with Old Testament scenes and themes) or New York beachfronts (perhaps alluding to the river and waterfronts Millet had peopled in his paintings), an artist's easel, something Steinberg shares with Millet.

The composition is held together with stable, serial, rectangular, and vertical units across the canvas. Several small "allusions" to portraiture infer a figure, again in different settings.

The *mode* is a variation on classical and romantic and really does not fit into either category or the ranges between the two.

The *Visual Form* is one picture unit, 32 inches high, takes on the illusion of importance by appearing to certify Millet's artistic life. In point of fact, the piece is whimsy. This interpretive biography is curious too, because Millet took his work seriously even though his critics did not.

Our next example is autobiographical. "Towards the end of my student days I had increasingly become dissatisfied with my subject matter and reasons for painting. I asked myself the question, 'What am I really concerned with at this particular moment?' In my second year at the Slade (a school of art in London, England), I had fallen in love with a beautiful girl, to whom I was returning from France in a few months to marry. This fact gave me the motivation and subject matter I needed—my family and its continuing story."[4]

This 1977 quote from the English artist Anthony Green was made at a time when "the narrative" was anathema to the world of art. But Green was purposive in his selection of "family" as subject matter. The everyday is exalted, even middle-class urban folk like his family.

The *Literary Form* is biography. The *Narrative Subject Matter* is his fourteenth anniversary. The *time frame* captures one moment in a monumental way so there is a timelessness about it, but particularized. His *sense about his own History* is "intensely funny and intensely true."[5]

The *Presentation Subject Matter* includes both himself and his wife inside a tent, at night, with an overhead light. *Materials* are oil paint on board. *Formal elements* do include a p.o.v. with this illustration. Our eye level is a little below the center and we are in a one-point perspective. There is little doubt about the center of interest because it is coincidental with the point of convergence of the receding lines on the horizon. Lines play an important role in this work. Lines of the tent canvas consistently bring the eye back to the center of interest. Values and textures align with realism. Colors are strong and sticky in appearance. The shapes and space here imply a life size scale, the painting size about 6' × 8'. Many of the shapes of the tent are triangular, leading one's eye back, along with lines, to the center of interest. The couple rests in a pyramidic unit. External lines of the canvas are pyramidic too, all of which helps give a sense of stability to the composition.

The *mode* here is classical with closed, known, and understood items. The *Visual form* is a large painting, 6' × 8', inviting us into an intimate moment in his life. We cannot move around Anthony Green. He is there with his wife, like it or not.

The photographic moment is caught beautifully by Michael Mazur's *The Incident At Fresh Pond* in a series of five panels, three of which we see here.

According to Mazur, a bucolic environment along with a rowdy group of kids reminded him of a former experience in another wooded area when he took part in a manhunt for killers of two local policemen. "Hundreds of armed and inexperienced people, without a clear idea for whom we were looking, rampaged the wood without system or caution in what was one of the most frightening experiences I had ever had."[6]

The *Literary Form* is autobiographical. The *Narrative Subject Matter* is a slice of his experience. His *sense about the autobiographic experience* is serious. His *time frame* is the immediate, journalistic moment in a series of five panels, three of which we see. The panels we see are about 3' high by 4 or 5' wide, so some of the figures could appear close to life

**Figure V.17** *Anthony Green. Our Tent. Fourteenth Wedding Anniversary. 1976. Oil on board, 72″ × 96″. Collection: Rochdale Art Gallery. [Seen in the* Narrative Paintings Catalogue, *Arnolfini/Clive Adams Exhibition. The Arts Council of Great Britain.]*

**Figure V.18** *Michael Mazur. The Incident At Fresh Pond. 1979. Detail. Pastel on paper. Three of five pieces. 40 1/2″ × 55″, 40 1/2″ × 48″, and 40 1/2″ × 55″. Barbara Krakow Gallery, Boston. Photo credit: Greg Heins.*

size. The *subject matter* includes running figures in a wooded area during daylight. The *materials* are pastels (chalks) on paper, which compared to the works of the same title in oil, feel a little more "breathless," as though we were part of the running group.

*Formal elements* begin with our p.o.v., which feels like a roving perspective as we move our eyes from left to right, sensing we are part of the group within the forest. Lines are blurred, predominantly vertical with diagonal slashings for tree limbs and leaves, plus some ground strokes. The patterns of light come from the sun's natural rays, which are somewhat behind the figures, filtering through leaves onto the figures and the ground. Textures are more implied through the "brushyness" of the stroked chalks in the vertical tree bark, the slanting limb, leaf, or needles of the trees, rather than explicitly stated through the drawing of actual textures. Shapes are underdefined, using suggestion or inference of runners both in foreground and background. Space carries a natural sense of recession. The composition is difficult to sustain in one panel alone. The sense of haste, of intensity, of all the goings-on in these works keeps one moving back and forth from one panel or another, experiencing the event as a participant. Because of the staccato lights and darks, the blurred and fuzzy "hurry" lines (which indicate the *mode* as romantic or gestural), the placement and body attitudes of the runners, we instinctively feel a sense of urgency within the composite composition.

The *Visual Form* is a tryptich here, three of five pieces, pastel on paper, framed, allowing us to feel the anxiety and frenzy of the hunt.

## A Study:

Kay-Lynne Johnson's autobiographical work is titled, *Family and Friends, Computer Self-Portrait.* Johnson moved into new artistic territories with laser drawings.

She selected specific childhood photographs, pictures of her family, and some words that characterized friendships or experiences. Using that autobiographical material she made several collages that she called "studies." The "stream of consciousness" (studied earlier in the paradigm on *Paradox*) took over as she manipulated her media, pushing and pulling space with oil pastels, oil crayons, pastels, gel medium, pencils, and torn bits of collage materials. The surface is very active, meaning that traditional Renaissance spatial illusion is not a factor. The work stays on or in front of the rectangular picture plane.

From the collages only parts were used, blown up, now transferred to the laser computer screen where she generated marks two ways. Some marks were applied directly to the surface; the others were achieved by guiding the laser itself.

Distortions that accidentally appeared served to alter human proportions. Those distorted shapes often came closer to her feeling responses than otherwise possible.

The *Literary Form* is *biography*. The *Narrative Subject Matter* is herself. Her *sense about The Story*, is seriousness with elements of chance surprises. The *time frame* indicates a compression, although linear time here is not important to this abstract work.

*The Presentation subject matter* involves people, friends, and events from her past. Different *materials* are used in two separate sequences. The first sequence was the making of the collage studies, as described above; the second step was using an opaque medium to make marks on the computer screen. Then she used the laser light medium to stroke over the opaque lines and shapes.

The *formal elements* indicate no particular point of view. Lines and values predominate. Diagonal lines move both left and right. A heavy **U**-shaped line below cradles a skull. Values often are dense, multi-layered, complex. Colors are muted into blacks, browns, golds, reds, greys, and whites. Textures are specific to the build up of "stuff" on the surface. Shapes are chance residuals, none held precious. Space is a dense, compacted, mysterious surface.

The *mode* is romantic. The *Visual Form* is one framed, computer-generated work, at once compelling yet removed, as computers so maddeningly are.

The *Comic Strip* is our last brief survey. It is included here because "comics" are narratives that are before our eyes daily and internationally. "Comics" are cited as a generic term to imply four things David Kunzle outlines in his article, "The Comic Strip."

1. There must be a sequence of distinct but causally interconnected images.
2. There must be a preponderance of image over text.
3. The medium in which the strip appears and for which it was originally intended must be reproductive, i.e. printed, a mass medium.

***Figure V.19*** *Kay-Lynne Johnson.* Family and Friends, Computer Self-Portrait. *1990. Laser computer generated drawing, Mixed media, computer collage. 21″ × 15″.*

4. The sequence must tell a story which is new, moral and topical, and which is couched in a popular idiom."[7]

The ancestors of the comic strip are pictures with subdivided image units as well as those that are serially interconnected. Both are offspring of the printing press and "broadsheets" is their name. We have already spoken about the predella paintings, small story-telling scenes surrounding or placed below a main altarpiece. Some of the earliest broadsheets were religious picture stories.

Following is one of the earliest surviving fragments of a German broadsheet.

The interpretation of this sequence would suggest the growth of the Soul. Men often consider their souls as feminine. Women do not cross-reference. Their souls seem to remain feminine. This image could represent either.

(a.)       (b.)       (c.)

(d.)       (e.)       (f.)       (g.)

(h.)       (i.)

***Figure V.20***   The Courtship of Jesus and the Christian Soul.
*c. 1470. A German Broadsheet. [Seen in the* ArtNews *Annual*
XXXVI, *1970,* Narrative Art, *Macmillan Company, N.Y.,*
*Newsweek, Inc.]*

a. *Jesus tries to wake the Soul, who wishes to stay asleep.*
b. *Jesus gives the Soul a love-potion.*
c. *The Soul seeks for Jesus, who hides behind a curtain.*
d. *Jesus tries to bribe the Soul to leave Him alone.*
e. *The Soul hears secrets from Jesus.*
f. *The Soul and Jesus are reunited.*
g. *Jesus entices the Soul with violin music.*
h. *Jesus disillusions the Soul with earthly things.*
i. *Jesus strips the soul naked.*

The Soul is closed, asleep. The Soul begins to understand unconditional love. She longs for more but knows not where to seek. Her burning heart wants only Jesus, who appears occasionally. Through continued longing in prayer, scripture, and action, the Soul hears the secrets of unconditional love. There is beauty in being together. The Soul is tempted to lesser things. She is stripped naked in the dark night of the soul, when the light is so bright one is blinded and one feels alone.

The *Literary Form* is parable. For the *Narrative Subject Matter*, the artist selected the growth of the soul. The *time frame* is sequential units, a compressed sequence of time in a person's life (these stages often take decades). The artist's *sense about The Story* is serious.

*Presentation subject matter* is the Soul with Jesus, seen in several settings. No special *light* source prevails. The scenes are high in value. The *materials* are ink and paper. The principle *formal element* is line. Some value is visible in the hair, but little else. Shapes are conceived exclusively as positive shapes. Space, interestingly enough shows early signs of linear perspective. You might remember that we spoke of the young Italian artist, Masaccio, who implemented consistent linear perspective in his paintings about

1425. This broadsheet is about 1470. The first panel, "a." a two-point perspective does not recess consistently. But within the "e." panel, the one-point perspective of the bench does. The *mode* would come under classical, known shapes.

The *Visual form* is a broadsheet. Step by step, picture by picture, one's soul grows.

The Swiss Rodolphe Topffer (1799–1846) is considered to be the father of the modern comic strip. He used a breed of "everyman" who struggles against life's caprices. And, as a draftsman, he varied physiognomies, moved away from careful rendering, inferred movement in his characters, and incorporated the broken, suggestive line.[8]

Religious monks think M. Vieux-Bois, "everyman," is from outer space. They imprison him, separating him from his lover. He attempts suicide. He changes his clothes, which always seems to make him feel better. He is depressed and falls sick. Finally he reaches his lover and rescues her.

Try applying the components of the Picture Form Chart to this comic sequence. See what you generate.

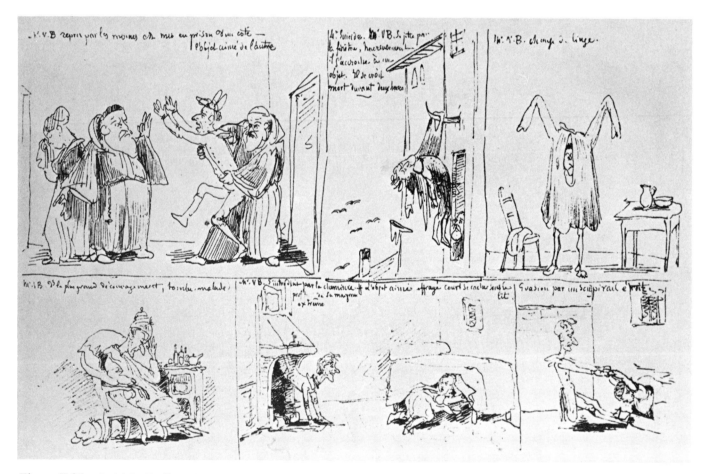

**Figure V.21** *Rodolphe Topffer.* Adventures of M. Vieux-Boix. 1827. *(Seen in* ArtNews Annual XXXVI, Narrative Art, *David Kunzle, "The Comic Strip," page 140.)*

## A Study:

The Todd Dykshorn comic book sequence that follows is about Freudian definitions personified.

Captain Ego rides the wave lengths of the conscious mind, the hero of our society and guardian of our morals. Accompanying Captain Ego are the snake, the Id, called Sid, and the Ego-Ideal. The three bonded together give Captain Ego the power needed to activate his defense mechanisms to carry out his infinite task—to suppress Eros, the female, (sexual desire) and Thanatos, the male, (the death wish). Eros and Thanatos are bickering siblings "forever attempting to exploit their drives upon a fragile society."

Captain Ego happens onto Eros and Thanatos, who immediately steals Ego-Ideal from him, leaving Captain Ego weak. Eros and Thanatos race back to their place in the Subconscious. They can now take over the city.

Meanwhile in society, the Morals, meeting in their doomed city, are unaware of their plight. Captain Ego is too disabled to heed their cry for help as Eros and Thanatos swoop down and capture the Morals. That leaves the two drives, Eros and Thanatos to reign. In the depths of the subconscious, the brother tells his sister of the Oedipus and Electra complexes which Sid, Captain Ego's only remaining influence, controls. Thanatos wants to persuade the snake, the Id named Sid, to come to his side.

Sid does! Captain Ego is now powerless and anguished. Thanatos wants complete power. But his irresponsibility worries Eros. Thanatos' greed motivates him to dispense with his sister by injecting her "realm of bodily pleasure" with his drive for destruction. This produces a Virus of Sexual Deviance to plague society.

All along, slithering around in the depths of the subconscious, the snake, the Id named Sid, having secretly remained loyal to his identity, at last finds his captured coparts and the Morals. Ego-Ideal, the Id and Captain Ego bond once more. Now Captain Ego is strong again and releases the Morals from the glass jar. Now he knows he must confront Thanatos and Eros.

But, down in the bottom right hand corner of the last picture board, Eros whispers to her brother, "My brother, don't dismay. I have another plan."

The *Literary Form* is fable. The *Narrative Subject Matter* is psychological fantasy. *Time frame* is a series of sequential units. The artist's *sense about The Story* is comedic. *Presentation subject matter* is a group of characters already identified. The settings vary from flying through space, to the city, to the glass jar, to the subconscious, and back to consciousness and personhood. Time is anytime. Lighting is strictly black or white. The *materials* are ink and board.

The *formal elements* most used are line and value polarities, (black and white), positive shapes, and space that bleeds from one situation into the next. The *mode* is classical, lines and known forms predominating.

The *Visual Form* is "Comics," with a distinct cause interconnecting the images, a preponderance of image over text, it can reach the masses, and it tells a topical, moral story which is couched in a popular idiom.

*Figure V.22*   *Todd Dykshorn.* Captain Ego. *1988. Ink on board, five story boards, each 40" × 30". Figure continued on next page.*

**Figure V.22**  *Continued* .

260

The reasoning process stressed in this paradigm may seem to be an after-the-fact process, one that applies once a work is completed. Actually, the process resembles the expressive, creative action itself, beginning with the narrative stimulus, then allowing you to move from choice to choice. Bit by bit, scene by scene, layer by layer, sequence by sequence. If you have ever worked on the rehearsal and preparation of a dramatic vehicle—a play or a musical, for example—you will appreciate the process. Your role as collaborator and interpreter gradually lets you reach a unique personal statement, a symbiosis, a coming together of word and image.

## ENDNOTES

1. John Ashbery. 1970. "Saul Steinberg, Callibiography," *Artnews Annual XXXVI*, ed. Thomas B. Hess and John Ashbery, New York: Macmillan, Newsweek, Inc., p. 53.

2. Pierre Gassier. *The Drawings of Goya.* New York: Harper and Row, p. 376.

3. Ashbery. "Saul Steinberg: Callibiography," p. 57.

4. Timothy Hyman. A quote from Anthony Greene, artist. From *Narrative Paintings Catalogue* at the Arts Council of Great Britain.

5. Ibid.

6. Michael Mazur. Nov./Dec. 1979. Frumkin Gallery, Hayden Gallery Catalogue, Massachusetts Institute of Technology, Cambridge, Mass.

7. David Kunzel, "The Comic Strip." *Art News Annual XXXVI Narrative Art*, 1970, New York: Macmillan, Newsweek, Incorporated, p. 133.

8. Ibid, p. 140.

## BIBLIOGRAPHY

Aichele, G. *The Limits of Story.* Philadelphia, Pa.: Fortress Press, 1985.

*American Narrative Painting Catalogue.* (Notes by Nancy Wall Moure.) Los Angeles County Museum of Art. Praeger Publishing Incorporated. (October–November, 1974).

*Artnews Annual XXXVI.* "Narrative Art." B. Hess and John Ashbery. New York: Macmillan, Newsweek Inc., 1970.

Ashbery, John. *Artnews Annual XXXVI.* "Saul Steinberg: Callibiography." New York: Macmillian, Newsweek Inc., 1970.

Austin, John. *How to Do Things with Words.* New York: Harvard University Press, 1965.

Bax, Dirk. *Hieronymus Bosch and Lucas Cranach: Two Last Judgement Triptychs, Description and Exposition.* Translated from the Dutch by M. A. Bax-Botha. Amsterdam and New York: North-Holland Publishing Company, 1983.

Bolt, Thomas. *Arts Magazine.* "Uneasy Answers: Lincoln Perry's Dialectical Narratives." November, 1986.

Brady, L. *Narrative Form in History and Fiction.* Princeton, N.Y.: Princeton University Press, 1970.

Bryson, Norman. *Word and Image: French Painting of the Ancient Regime.* Cambridge and New York: Cambridge University Press, 1981.

Cook, Albert Spaulding. *Changing the Signs: The Fifteenth-Century Breakthrough.* Lincoln: University of Nebraska, 1985.

Crawford, Hubert H. *Crawford's Encyclopedia of Comic Books.* Middle Village, N.Y.: Jonathan David Publishers, 1978.

Daniels, Les. *Comix: A History of Comic Books in America.* New York: Outerbridge and Dienstfrey; Distributed by E. P. Dutton, 1971.

Eaton, Tina and Shoen, Mary. *Mark Van Proyen Catalogue.* Shoen Gallery, San Francisco, Calif., April, 1981.

Edgerton, Samuel Y. *Pictures and Punishment: Art and Criminal Prosecution During the Florentine Renaissance.* Ithaca, N.Y.: Cornell University Press, 1985.

Fagaly, William. *Southern Fictions Catalogue.* Houston, Tex.: Contemporary Arts Museum, (August–September, 1983).

Fermigier, Andre. *Jean-Francois Millet.* Skira, New York: Rizzoli International Publications, Inc., 1977.

Flaxman, Rhonda Leven. *Victorian Word-Painting and Narrative: Toward the Blending of Genres.* Ann Arbor, Mich.: U.M.I., Research Press, 1987.

Fry, Phillip and Poulos, Ted. *Steranko: Graphic Narrative Catalogue.* "Storytelling in the Comics and the Visual Novel." The Winnepeg Art Gallery, Canada, 1978.

Gaballa, G. A. *Narrative in Egyptian Art.* Verlag Phillip at Vin Zabern, Maine, 1976.

Gassier, Pierre. *The Drawings of Goya.* New York: Harper and Row, 1975.

Glowen, Ron. "Uses of the Figure." *Artweek.* (March 1, 1986).

Goldin, Amy. "Words in Pictures," *Artnews Annual XXXVI*, New York: Macmillan, Newsweek, Inc., 1970.

Goodman, Susan Tumarkin. *Jewish Themes Catalogue.* Jewish Museum, New York.

Hayden Gallery Catalogue "Michael Mazur," Massachusetts Institute of Technology, Cambridge, Mass., Nov.–Dec. 1979.

Hollander, John. "Crimes of the Art." *Art in America*, (April, 1986).

Horn, Maurice. *The World Encyclopedia of Comics. Volume I and II.* New York: Chelsea House Publishers, 1976.

Hyman, Timothy. *Narrative Paintings Catalogue.* Arts Council of Great Britain, 1980.

Johnson, Edward Dudley Hume. *Paintings of the British Social Scene, From Hogarth to Sickert.* London: Weidenfold and Nicolson, 1986.

Kardon, Janet. *Image Scavengers: Painting Catalogue.* The Institute of Contemporary Art, University of Pennsylvania. January, 1983.

Keach, William. *Elizabethan Erotic Narratives: Irony and Pathos in the Ovidian Poetry of Shakespeare, Marlowe, and Their Contemporaries.* New Brunswick, N.J.: Rutgers University Press, 1977.

Kunzle, David. "The Comic Strip."*Artnews Annual XXXVI.* New York: Macmillian, Newsweek, Inc., 1970.

Kuspit, Donald. "Of Art and Language." *Art Forum.* May, 1986.

Leitch, Thomas M. *What Stories Are: Narrative Theory and Interpretation.* University Park: The Pennsylvania State University Press, 1986.

Lucie-Smith, Edward. *Work and Struggle,* "The Painter As Witness, 1870–1914." London and New York: Paddington Press, Limited and Grosset and Dunlap.

Male, Emile. *Religious Art in France: The Late Middle Ages; A Study of Medieval Iconography and It's Sources.* Edited by Harry Bober; translated by Marthiel Matthews. Princeton, N.J.: Princeton University Press, 1986.

Marable, Darwin. "Messages From the Id." *Artweek.* January 18, 1986.

Marincola, Paula. *Words and Images.* Philadelphia College of Art Catalogue, (September/October, 1979).

Mellard, James M. *Doing Tropology: Analysis of Narrative Discourse.* Urbana: University of Illinois Press, 1987.

Nicholsen, Chuck. "To Amuse, Bewilder, Annoy and Inspire." *Artweek.* February 8, 1986.

Okudaira, Hideo. *Narrative Picture Scrolls.* Translation adapted with an introduction by Elizabethen Grotenhuis. New York: Weatherhill, 1973.

Owens, Craig. "Telling Stories," *Art in America,* May, 1981.

Polanyi, Livia. *Telling the American Story: A Sculptural and Cultural Analysis of Conversational Storytelling.* Norwood, N.J.: Ablex Publications Corporation, 1985.

Prince, Gerald. *Narratology: The Form and Functioning of Narrative.* Berlin, New York, Amersterdam: Mouton Publishers, 1982.

Prince, Gerald. *A Dictionary of Narrotology.* Lincoln, Neb.: University of Nebraska Press, 1987.

Ratcliff, Carter. "Illustration and Allegory," Catalogue, (May–June, 1980). New York: Brooke Alexander, Inc.

Rigberg, Lynn R. "Earl Linderman." *Arts Magazine 55,* March, 1981.

Rosenzweig, Phillis D. *Directions, 1983 Catalogue.* Hirschorn Museum and Sculpture Garden. Smithsonian Institute, Washington, D.C., 1983.

Sartin, Stephen. *A Dictionary of British Narrative Painters.* Leigh-On-Sea: F. Lewis, 1978.

Shannan, Joe. *R.B. Kitaj Catalogue.* Hirschorn Museum and Sculpture Garden, Smithsonian Institute, Washington, D.C., 1981.

Stewart, Susan. *On Longing: Narratives of the Miniature, the Gigantic, the Souvenir, the Collection.* Baltimore: Johns Hopkins Press, 1984.

Wollheim, Richard. *Art and Its Objects.* New York: Cambridge University Press, 1980.

Young, Katherine Galloway. *Taleworlds and Story Realms.* Boston: Martinus Hijhoff Publishers, 1987.

# Putting It Together

▼

When all is said and done, when all the exercises, the experimenting, philosophizing, starting over and looking, testing, and sharing are processed, what makes up an informed image will shine through. And not everyone will approach each paradigm with the same amount of enthusiasm. Approaching opportunities in them or around them is what matters. Being honest, open, and willing to change allows a mind-set for creative solutions.

Each paradigm stresses one or more issues every artist could master to find and retain a personal vision. Forming is a requirement, of course, while proficiency guards against purely random but ignorant solutions. However, randomness is a kind of readiness, which should infer a working confidence, a capacity for "seizing the moment." Those moments will be empty, however, without a sense of visual architecture, or without owning *your* iconography or symbolic images.

Work of the ensuing students each reflects a process, a breakthrough, if you will. These works range from realism to abstraction and many degrees in between. Drawing the figure to me means both *drawing the figure* with the eye and *drawing about the figure* from within. Finding who you are through your stroke is a constantly fascinating two-way journey.

## Peter Hansen

The ideas for my work come in odd bits and at moments like sitting and thinking on a bus, taking a break in a conversation, following a fleeting whim, a childhood memory, or a smart remark. My brother, my twin, said one day, "I wonder if you'd have to pay twice at the box office if you were a Siamese twin." That remark spun me into the whole series of works using associations between twins.

The way I work is simply to make layers of gestural marks at first. I try not to think too much except to think of something mischievous; otherwise the lines become labored. Mental images are most important. Color decisions are done spontaneously and quickly while I go for the anecdote. The words are a point of reference, a nonsense, an emotional response to what's been done.

I try to balance humor/ seriousness, the childlike/childish. The message rests as it is. I call myself an Emotional Realist because my images are most real to me. My work is finished when my energy becomes labor.

Together, of course, they conjured up rotten spirits for evil THIEVES of sacred arts. *1989. Pastels on rag paper, 30″ × 20″.*

*Kristin Lucas*

They Are Me. *1989. Chalk and oil pastels on rag paper, 20″ ×
30″.*

**M**y imagery comes from my innermost thoughts, my dreams, my poetry. Most of my images do not include words because right now I find the images say the same things. I am beginning to incorporate words. But these pieces show past feelings, recurrent feelings, from patterns in my life. I distance myself from problems, and feel I am in a magical dream forest. My work belongs to that forest.

I use chalk, oil pastels, and rag papers. Because the strokes belong to me there are no mistakes. I try to combine image and background, and in representing feelings, I choose colors. To express tension for a split personality I use reds. I chose green for the foreground, the complementary color to red, the opposite, thus denoting the split personality. The images, the colors, are a self-portrait, they are me.

When I sense I have become the emotion I am representing, the work is done. However, I am never comfortable with a piece. It will propel me to do another.

*Brent Dierks*

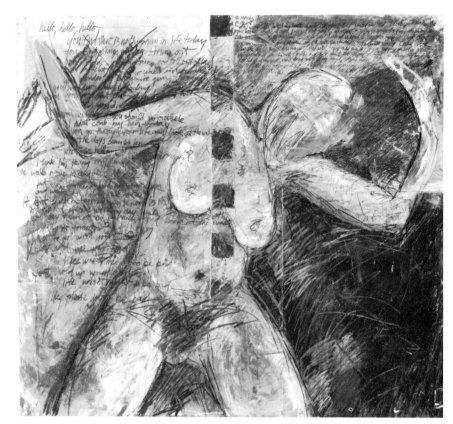

Of Me to Someone[2]. *1989. Old drawings, paste, gesso, pastels, matte medium, acrylic sprays on masonite, 5′ × 5′.*

**M**y ideas come from human relationships. I think about relationships, then try to work things out. My drawings are therapy. The writing is not an essential part of the image, but it needs to be there, interplaying with the figure. It goes along with other symbols. The writing is a talking back and forth about what is going on, but it does not overpower the image. The writing ties it together. The writing invites the viewer in and makes the work immediately more personal.

I started with the male/female relationship and have gone on to other relationships like the +/−, the x/o's, black/white and from relationships of two to relationships of several, but the relationships are always of me to something or someone.

I use masonite, sand it down, then gesso the surface. I use old drawings from life-drawing class and paste them onto the masonite with gesso and then work up a surface on which to draw. I beat the heck out of the surface. With this particular work I outlined the model's body in pencil within the form of a square, which feels more stable to me. A horizontal format pulls me over and out while a vertical format pulls me up and out. I feel more comfortable, too, using the main torso within the square, not including arms and legs.

Once I get the composition, I go back in with pastels, matte medium, and acrylic sprays to set things in place.

When the image making is done, I use an acrylic gloss so that light can create richer colors.

When I begin to draw, the piece changes every 5–10 minutes. The more I work, the more slowly the changes come. When I think I am done, I sit in a chair about 10 feet away and stare at the drawing for one-half hour. I divide the piece into sections and see if all is working in that one area. If it is, I'll leave it. Then I look at that section in relation to the whole piece. If that works, I go to another section and do the same. And if all sections work, it is done. I check day to day. When I apply the last spray coating I won't touch the piece again.

*Ken Wong*

Neanderthal. *1989. Collage of litho crayon, photocopies, and glue, 20" × 30".*

The images come from the chance strokings on a sheet of paper. I look for images, then try to create a theme. The process cannot be separated from the image. I use litho crayon, a photocopier, collage, erasers, scissors, and glue. I take some drawings, half done and photocopy parts of them. I change the scale making some images larger or smaller. Photocopying makes the strokes flatter, less dense, and gives them a different tone. Sometimes frottage, the stroking of media on a paper placed over a rough surface, is used for textures.

The theme in this one involves TV, past and present. The work includes references to Neanderthal man, with cave paintings on the right. They are in contrast to technology with satellites above Ernie Kovaks' face. On the left side is an African mask, which could relate to Walt Disney or Marlin Perkins. The gorilla on the bottom is shooting out death rays, another reference to Ernie Kovaks or the Nairobi trio. The woman is shielding her face.

The meaning could be Ernie Kovacs coming back from another world for syndication, like a savior. It could be a tribute to his genius. Or it could be a commentary on technology.

If I can get more stuff onto an image and have it work, I know I am not done. It has to be a controlled chaos.

## Peter Rian

The idea for this work came from a critique in a lithography class. There were two pieces next to each other on the wall that seemed interesting. The rest of the work in the class was very weak. Students would say things like, "Well, the composition is so settled." They were trying to attach words to the pieces using talk as filler, with no one aware they were treading time, like water. These students created works using formal elements, line, value, and such. Then they attached emotions to that. None of them seemed to understand that formal qualities don't make art, Art. Emotions need to get into the work. Their words and works seemed shallow. I was really angry over the class. No one took me seriously.

That evening I began drawing, still angry over the earlier class. I did not know what was going to be drawn. I felt I wanted to make a fragile but heavy image, a thin facade that is hard to see through. I recalled the two drawings that seemed to have merit and began drawing them. Each of the other small works drawn emphasized one formal element like value or line. I wanted to illustrate that when skill becomes the work instead of skill being used as a means, the ends are superficial.

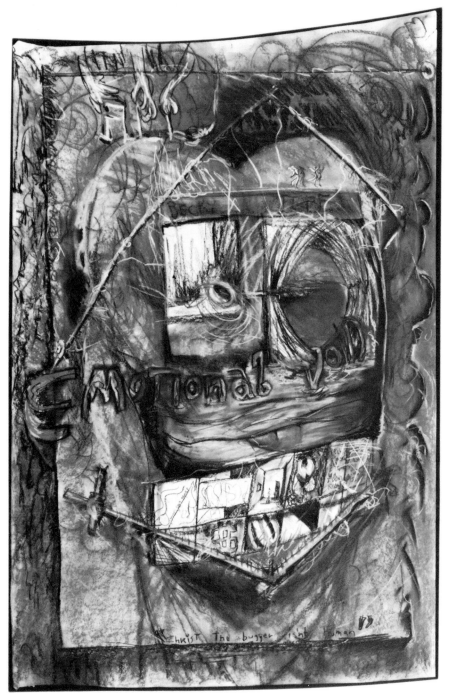

Fill That Emotional Void. *1989. Pastels and drawings applied to black paper, 4′ × 3′.*

I developed this idea as I worked on it, using a 4′ × 3′ black paper with charcoal sketches on first. I chose which marks were to stay, then added colors, pinks, blues, reds, green, and browns at the last.

This piece began for a specific audience that was unaware of an emptiness, of the air inside themselves. I wanted them to have an idea of what art could be, what it is. Otherwise, the intimacy with art is lost.

My work is finished when it has no room for anything else. I know I am finished at the same time the work is finished. Sometimes I am finished and the work isn't, then I can never complete it.

## Pam Shewanick

My ideas usually come late at night, between midnight and 4 A.M., when I am having coffee by myself. The images represent my responses from something I have read in the newspapers or seen in the newscasts. But the images are not narrative. They reflect feelings about issues, but not the issues themselves. I try to move to an issue to incite an intense response, and out of my frustrations the work takes on a destructiveness. As the figures are, so is the world: crumbling, dismembered, detached.

I use paper, oil pastels, chalk, pencils, colored pencils, and graphite. I begin with borders, like a frame. I do not like empty edges. Then I just start drawing. I am not hesitant about putting marks down. Layering for me is putting on colors and scratching back into them. Backgrounds are space, lines or movement. The images come, they are not thought out.

My work is finished when I do the whole thing in one shot. If the piece cannot be completed, it tends to go on forever. When I can't finish something, I get upset, wad the paper up, and stick it in a corner.

I Just Got Out of the Shower. *1989. Lithograph 6" × 16".*

269

*Richard Heger*

The Step-Mother. *1989. Charcoal and white conté on charcoal paper, 24″ × 18″.*

I draw what is there. This person seemed distant and detached. I was fortunate in that I drew her in about 1-1/2 hours while she was watching TV. She does not enjoy being observed. I used charcoal, white conte, and charcoal paper. I began gently, loosely, gesturally allowing the composition to develop in the drawing process. Once the composition was secure, I worked on the eyes because I feel that is the most essential part of the person. Then I worked on other areas trying not to think about the eyes because they are too distracting. I used the charcoal to cover large areas lightly, then covered the next darkest shapes until I felt the work was finished.

The work is done when I feel it is compositionally sound. I know the piece is finished when I can look at it without feeling something is wrong.

*Scott Jepsen*

Forbidden Image. *1989. Conté crayon and oil pastel, 24″ × 18″.*

The imagery for my work comes out of the shapes I find in the mass of scribbles on the paper after I have worked over the surface for a while. I call these works *Forbidden Images*, because they are hidden in the psyche and I cannot touch them logically. These are subconscious images, less refined and more spontaneous than works that are done realistically. I find my own stroke in the play of drawing.

I use conte crayon, and oil pastel and begin making marks drawing from a model within her environment. I draw ten to twelve gestural drawings of sixty seconds each, one on top of the other. I smear the marks, erase the images I don't want, keep the ones I do want and then add color onto the shapes. Additional black marks help shift or tighten the composition, but I try to retain a sense of playfulness with the media.

My work is finished when I feel it does not need anything else. I leave a work, go back to it, rework it and sometimes create an entirely new piece.

## Paul Guy

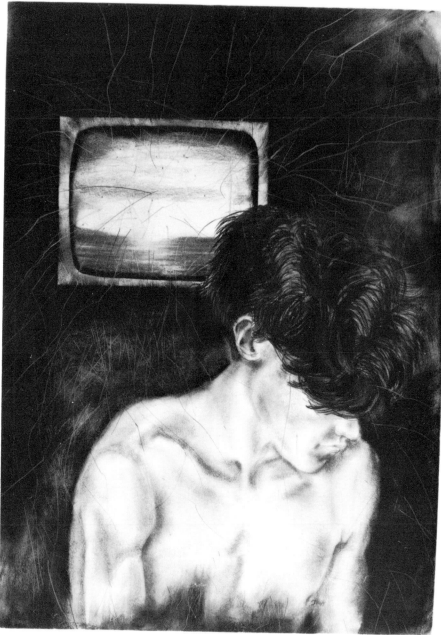

Self Portrait. *1989. Charcoal, carbon, pencil on rag paper, 30″ × 20″.*

The ideas for my work are always conscious. I have to work at them.

Little side projects quite apart from my drawings become starting points. Once I buried a dead bird and documented that with photos. The photos became a starting point for me. Or I build environmental sculptures from found twigs, rocks, and sticks, then let those sculptures prompt ideas for my drawings. Another process I use is trying to capture the feelings of conflicts between other people and myself. I work at manipulating ideas into images, working out all the formal things, lines, colors, and so on before I start.

The materials I use are charcoal, carbon, pencil, and Rives paper. I begin in an irrational way making marks on a surface, scraping in a haphazard way that feels like making "static" lighting effects. I build and build the surface, working the whole drawing all the time, and finally completing the image that was planned.

I let a work sit around for a few days, come back, look at it, go back into the drawing, then put it away before I feel it's done. I return to the drawing for confirmation, then leave it alone.

*Laurie Cinotto*

Paul's Testament. *1990. Gesso, paste, acrylic, charcoal and pastels on four sheets of rag paper, 5' × 3½'.*

Τhis drawing was the first of a series of four large works, each five by three and one-half feet. The series was titled *Paul's Testament.* One panel is shown here.

I began by piecing together four sheets of Arches coverstock, tacking them to a wall, then coating the surfaces with a layer of gesso mixed with modeling paste. About this stage, "the idea" began its development, not from outside sources but from introspective roots. Returning to the surface, acrylic paint, modeling paste, charcoal, and pastels were layered one over another. Colors remained dark with warm blacks, umbers, and siennas. Media was applied with brushes and with my hands. Spraying the surface with alcohol and water allowed pigments to run and drip. This process also softened the charcoal so it smeared easily into the cracks of the veneer.

Each time I worked on this piece I covered the surface with a new layer of media, strokes, and drippings. I feared a certain amount of inconsistency if I chose to work selected areas of the piece. A distinct kind of depth is created by working this way. Because my work embodies a variety of mood swings, depth can be found two ways: the first rests in the physical appearance of layered media, the second is imbued within the content of the work, the psychological space.

When the piece neared completion, I began scratching away at the surface with common nails, etching needles, and my fingernails. With the last scratches, the work felt complete. Finally, the four large sheets of paper were recessed into a four-inch frame to support the weight of the papers and to help the work appear more solid.

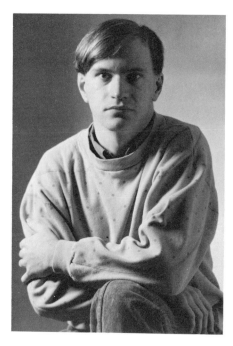

Nike. *1989. Mixed media, 22" × 31".*

*Jim Negley*

In my work, my subjects/themes are developed from things in life that catch my attention. I see objects when I am driving in the country or around town. With a few of those objects something clicks. Often the item appears to be a contradiction such as extremely tall radio antenna vertical to the ground that did not blow over. I think about objects in different contexts with different proportions, huge antenna in tiny landscapes, for instance. Almost unconsciously I establish a modular proportion within a work using the length of the colored antenna segments. Looking out a window

I generalize masses. The tops of trees and the volumes of the trees are translated into horizontal bands that become a landscape. I just use the green shape. If I used blue instead of green, the same shapes could become sky.

The Nike of Samothrace invites another contradiction I often use in my clay pieces and that is cantilevering a huge amount of clay into one area. I like the tension that creates. And, an hourglass form can represent itself or a woman's torso.

Once I get some ideas in my head I think about what unifies them all. I follow them like branches to the trunk

of a tree. I sketch these ideas, adding and subtracting until I discover something. Then I draw.

The drawing begins as a light pencil or charcoal gestural drawing on a sheet of paper, modified until the composition is right. Broad strokes of thin acrylic paint are applied to the large general masses. Over the underpainting layers of conte, pastels, charcoal, and other dry media define and finalize the idea. The work is finished when I feel there is nothing left to draw.

*Paul Hempe*

Adav and Eem. *1989. Mixed media.*

**M**y ideas come from two sources, life and pre-Renaissance works of art. I like the funny, quirky things in illuminated manuscripts and want to play with their ideas and religious anecdotes such as imagining the figures in different contexts. My work is states of being or layers of self-portraiture, either spiritual, psychological, or physical.

The materials used are colored ink and gouache. The drawing process is working out the composition in line over layers of colored ink. About 85 percent of the time I begin with the title and a label. The label is illustrated through an image, as in *Adav and Eem*, a play, of course, on Adam and Eve.

The ideas are in my head, then to the sketch pad. I make notes. I combine shards of information, making a series of sketches for scale, watching for accidentals, and waiting for the finalized composition to appear. The combined images need to work together. Sometimes there are four to seven complete sketches, each done on a fresh sheet of paper before the work is ready to finalize on the final sheet of white paper.

A certain amount of perfection, idealization, and intellectualization needs to be modified to retain a sense of lightheartedness. Opulence and dense circles are needed. Blues, blacks, magentas, yellows, greens, gold, and silver are applied to retain playfulness and spontaneity. When the space is filled with enough stuff, I know the work is done.

## Lu Bro

My current work begins with 30 × 20″ sheets of drawing paper sized with matte varnish applied with a brush of heavy bristles that textures the surface. Oil paint is then spread in a variety of ways—in broad areas and in thin lines, one color over another. Often oil-based crayons are used. The method is random and only skeletal judgements about color lead me to pick and choose where to place what. I work to a similar stage about four pieces at a time, using five or more colors on each.

Several days of drying yield a tacky surface after turpentine is applied. Excess paint is rubbed off with paper towels. The large sheets of rag paper are rotated 90 degrees several times. From the remaining paint, latent images are scanned for possibilities. If one image looks promising and seems to be pivotal for the whole work, that image is "pushed," and bought to completion.

The delight in working this way is the "Aha" experience when something unexpected arrives and plants itself into the paint unforgettably. Such was the arrival of *Grandfather*. No doubt, a personal icon, the figure was too good to let go. I felt he took life seriously enough to have fun with it.

Grandfather Always Dressed for Miracles. *1985. Oil on sized rag paper, 30″ × 20″.*
*Collection of the artist.*